More LTD Stirling Engines

You Can Build

Without a Machine Shop

By Jim R. Larsen

ISBN-13: 978-1523667147

ISBN-10: 1523667141

Published by:
Jim R. Larsen
PO Box 813
Olympia, WA 98507
USA

Contact Information:

Email
Jim@StirlingBuilder.com

Website
http://www.stirlingbuilder.com/

Blog
http://woodenmusic.blogspot.com/

Twitter
@StirlingBuilder

YouTube
https://www.youtube.com/user/16strings

Contents

Introduction

Low temperature differential (LTD) Stirling engines are hot air engines that are highly efficient. Hot air engines operate when heat is applied to one side of the engine and the other side is allowed to cool. The difference between the temperatures of the warm side and the cool side is known as the "temperature differential." Engines that will operate with very small temperature differentials are labeled "LTD", or "low temperature differential" engines.

There is no current standard for exactly what the threshold is to earn the label "LTD", which has led to some confusion. For the purpose of this book, LTD engines are defined as follows:

The "low temperature differential" label applies when a hot air engine has these characteristics:

1. The engine does not require the heat of an open flame to operate.
2. The engine will operate on the temperature differential created by hot, non-boiling water in a 70° F (21° C) environment, at or near sea level.
3. The engine does not require ice or other cooling agents that are created or maintained by another energy source.

The temperature differentials shown for the engines in this book are determined by measuring the surface temperatures of a working engine. Those numbers are significantly different than the temperatures of both the heat source and the cooling agent. Using surface measurements on the engine and the standard listed above, the upper limit for the temperature differential of an LTD engine is about 100° on the Fahrenheit scale (55.5° C).

Engines that can function on a temperature differential of 40° F (22° C) or less are often capable of operating on solar energy using un-concentrated sunshine. If the engine will operate on a temperature differential of 20° F (11° C) or less, it can usually be made to run for extended periods of time using warm water as the heat source. Engines that will operate at a temperature differential of 10° F (5.5° C) or less will run from the heat of a warm hand in an environment that is 70° F (21° C) or less.

This definition covers a wide range of operating conditions, all of which fall in the category of "Low Temperature Differential." By contrast, High Temperature engines are often run using concentrated sunlight or flames that are hot enough to make metal parts glow red from heat. These engines run under very different conditions, and often have very little in common with LTD designs.

The name "Stirling Engine" was originally used to describe hot air engines that utilized the "Economizer" or "Regenerator" as patented by the Reverend Dr. Robert Stirling in 1816. His innovation increased the output of the hot air engine so that it became a viable working motor. Since then, "Stirling Engine" has become a broader term used to describe many types of hot air, external combustion, temperature differential motors, even if they do not use the features of the 1816 patent.

Reverend Stirling's "Economizer" is a physical structure situated in the airflow between the hot and cool sides of the engine. The Economizer captures and holds some of the heat from the air before it enters the cool side of the engine. This heat is then used to pre-heat the air as it flows back into the warm side of the engine. This has the effect of recycling some of the energy that would otherwise be radiated out through the cool side of the engine.

The engines illustrated in this book are described as "Stirling Engines" in the broader sense of the term, as they do not feature an Economizer.

Can These Engines be Scaled Up?

One of the questions frequently asked about small Stirling engines is if they can be scaled up to do real work. The short answer is yes, these engines can be scaled up to do meaningful work, but that answer comes with many conditions. While the Stirling engine is a champion of efficiency, it is generally not as powerful as internal combustion engines. There is a YouTube video of an LTD engine scaled up to support a small woodworking operation. The engine uses a large greenhouse to collect solar energy for the engine and produces enough power to drive power take-off (PTO) tools or a single electric hand tool.

There are Stirling engines in the commercial market that have been designed specifically for power applications. These engines are quite different in design from the engines shown here. They usually operate with hydrogen or helium inside instead of air, and the mechanics of the engines are so radically different that you might not think they are related when you compare them side by side.

If your goal is to build a Stirling engine as a power source for your home, cabin, or vehicle, you have an admirable goal. Building the engines in this book will teach you a great deal about LTD engine behavior and design, but these little engines will not power your home.

Stirling engines are more efficient than internal combustion methods, but they are less powerful. The graph on the next page illustrates how the two methods compare. The power curve for the internal combustion engine shows that power continues to increase as more fuel is added. This curve eventually flattens out when the internal combustion engine has reached its maximum output. (The graph is for comparison only and is not based on calculated engine performances.)

The Stirling engine is slightly more efficient in low power applications. But the power curve on the Stirling engine flattens out much sooner than the internal combustion engine. Once the engine reaches its peak power output there is nothing to be gained by adding more fuel. In some cases the output of the Stirling engine is likely to decrease if too much heat is applied. If heat enters the engine faster than it can escape, the engine overheats and slows down.

The principle reason the Stirling power curve becomes flat is because the air inside the engine has a finite potential for expansion. Once it has expanded to its full potential it will not expand further. Internal combustion is different in that adding more fuel can create a more powerful explosion.

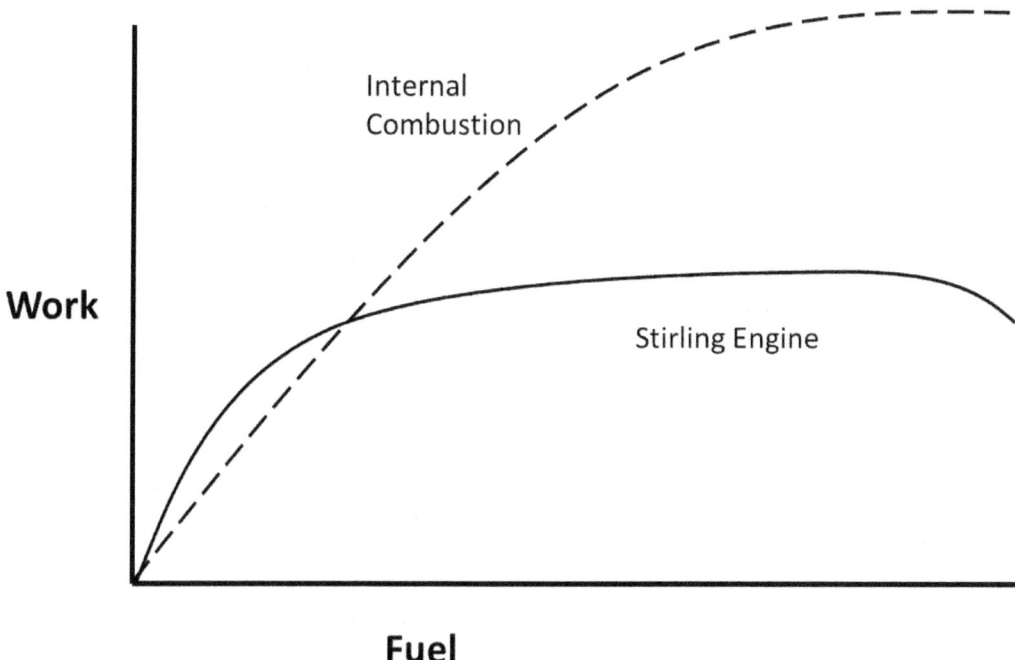

This graph illustrates the difference between the power curve of a Stirling engine and an internal combustion engine. The Stirling engine is more efficient in low power applications while being less powerful overall.

This fact can be mitigated in Stirling engine design by replacing the air inside the engine with another gas that is lighter than air. Lighter-than-air gasses have a greater potential for expansion and contraction as the temperature changes. The gas that is chosen for the interior of the Stirling engine is referred to as the "working fluid" of the motor. The most popular choices for increasing power output are helium and hydrogen.

Helium is affordable and easy to obtain in the commercialized regions of the world. The engines in the previous Jim Larsen book, "Three LTD Stirling Engines You Can Build Without a Machine Shop" were all designed with helium operation as an option. Adding helium to those engines creates an increase in engine speed of about 10% to 15%. Those engines all have a sealed pressure chamber and a magnetic drive mechanism. The effects of a helium charge can last up to 2 weeks.

The engines in this book can be equipped with vents for adding helium, but the effects will not last as long. The pressure chambers of the engines in this book are not sealed units. Each pressure chamber in these designs has a shaft that passes through a pressure chamber wall to control the position of the displacer. Any helium added to the engine will eventually leak out through the joint where the shaft passes into the engine.

Hydrogen is more challenging to work with. The hydrogen molecule is so small that it will pass through most metals. It doesn't leak through very fast, but it does leak. Because of this, Stirling engines that use hydrogen as a working fluid usually have a hydrogen generator integrated into the motor. The motor uses electrolysis to create hydrogen from a small water reservoir that must occasionally be refilled on the motor.

One additional step that can be taken to increase the output of a Stirling engine is to pressurize the motor. Adding more working fluid means there is more stuff inside, expanding and contracting as the engine operates. There are multiple pressurization schemes in use. Some involve an engine design with opposing cylinders that allows the internal workings of the engine to be pressurized. Another method is to place the entire engine in a pressure chamber and increase the pressure of the operating environment of the engine.

Since atmosphere thins as altitude increases, a non-pressurized Stirling engine will have decreased performance at higher elevations. This is because there is less air (reduced atmospheric density) to do the work of expanding and contracting inside the engine.

The engine designs illustrated in this book are based on what is referred to as a traditional "pancake style" pressure chamber that has an internal displacer moving up and down inside the engine. One of the more famous engines that uses this configuration is the LTD engine developed by James Senft in his work as a NASA researcher. James Senft has authored an informative series of books about LTD Stirling engines that are good additions to the library of any Stirling engine enthusiast. Some of his books include plans for building the motors, but will require machine tools such as a lathe or mill to complete the projects.

LTD Engine Designs in this Collection

The four engines featured in this book were selected from over a dozen different variations of LTD Stirling engine designs. These engine designs were chosen for their elegant simplicity and ease of assembly. Some might view this as four variations on a single design, which may indeed be the case.

The engines in this collection are numbered #4, #5, #6, and #7. Engines #1, #2, and #3 are in the book, "Three LTD Engines You Can Build Without a Machine Shop", also by Jim Larsen.

The larger engines in this collection are the 6" (15.24 cm) engines. The smaller engines measure 4" (10.16 cm) across the surface of the pressure chamber. Plans are provided for both a traditional round pressure chamber, and a simpler square pressure chamber.

All of these engines feature a drive assembly mounted on a central pedestal. This approach to the drive mechanism design makes it possible to avoid the tedious task of bending a crankshaft. This design also has fewer friction points than a crankshaft design would have, which is an added benefit.

Why Build a Square Engine?
People familiar with LTD Stirling engines often expect them to be round. The round engine design has been around so long, in fact, that many assume it is designed that way to somehow enhance its operation. The reason that the traditional design is usually round is because the LTD Stirling engine has for many years been a project built by lathe owners. The metal lathe provides the ability to make cylindrical objects with great precision. Many home hobby enthusiasts with machine tools have a metal lathe in their collection of tools. The LTD Stirling engine requires a high degree of precision to make well, and the lathe is a perfect choice for making pistons, cylinders, and bushings. Because the lathe can also be used to machine the top and bottom plates of the pressure chamber, they were made round as a result of the tool chosen to make them. The

round shape of the traditional LTD Stirling engine is not due to any increase in performance, although it could be argued that it is aesthetically pleasing to look at.

There are several advantages to building a Stirling engine with a square pressure chamber. The hot and cold surfaces of a square engine have a greater surface area than a circle of the same diameter. The greater surface area means that there is a larger heating surface to warm the air, and a larger cooling surface to chill the air inside the pressure chamber. The larger surface area means that there is also a greater volume of working fluid (air) inside the engine. A higher volume of air, and larger surfaces for heating and cooling that air, translate into a more powerful and efficient engine.

The square engine is much easier for a home builder to construct, especially if the home builder is working without the benefit of expensive machine tools. The square engines in this book utilize straight rectangular pieces of acrylic to make the pressure chamber sidewalls. These parts can be made with simple straight cuts and do not require any complex process to bend them into a circle.

Teflon tubing replaces expensive ball bearings. A drive diaphragm made from a rubber glove replaces the graphite piston and ground glass drive cylinder of a traditional LTD engine. The precision of the metal lathe is replaced by the craftsmanship of the builder and a meticulous attention to detail. All of these factors work together to create an LTD Stirling engine design that can be built with common hand tools, or with a few power tools frequently found in a home shop.

How These Hot Air Engines Work

The pressure chamber contains a small amount of air that is held captive inside the engine. One side of the engine is warmed, and the other side is kept cool. In the following illustration the heat is applied to the bottom of the pressure chamber and the top is not heated. A large loose fitting piston called the "displacer" moves the air inside the engine, from warm, to cold, and to warm again in a repeating cycle.

The air inside the engine expands when it gets warm, and pushes outward on the drive mechanism. When this same air is moved to the cool side of the engine, it contracts. This pulls in on the drive mechanism.

The drive mechanism pushes and pulls on the crankshaft. This causes the crankshaft and flywheel to rotate. The rotation of the crankshaft causes the displacer to rise and fall inside the pressure chamber.

The rotating flywheel stores enough energy to keep the crankshaft turning and cause the cycle to repeat. The air heats, expands, and pushes the crankshaft through the expansion phase. This moves the displacer and causes the air to enter the cool side of the engine. The air cools, contracts, and pulls the crankshaft through the contraction phase. This starts the next expansion phase and the pattern continues to repeat itself.

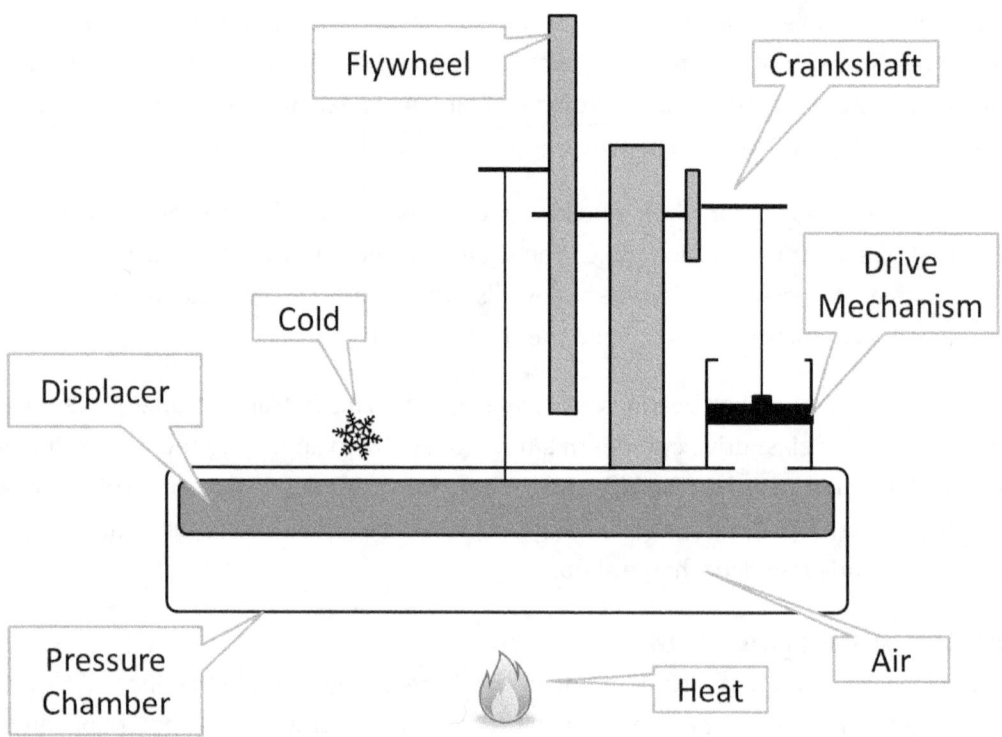

Stirling Engine Parts – This illustration shows the basic parts of an LTD Stirling engine. The pressure chamber contains air that is held captive in a closed system. The air is heated when it is on the bottom of the pressure chamber, which is exposed to a heat source. The air is cooled when it is on the top of the pressure chamber. The air is moved about inside the pressure chamber by a loose fitting piston called a displacer.

The flywheel stores enough kinetic energy to keep the engine moving smoothly between power strokes. The shape of the crankshaft keeps the motion of the displacer and the drive mechanism synchronized. They are synchronized with a 90° phase angle between them. The drive mechanism is always a quarter turn behind the motion of the displacer.

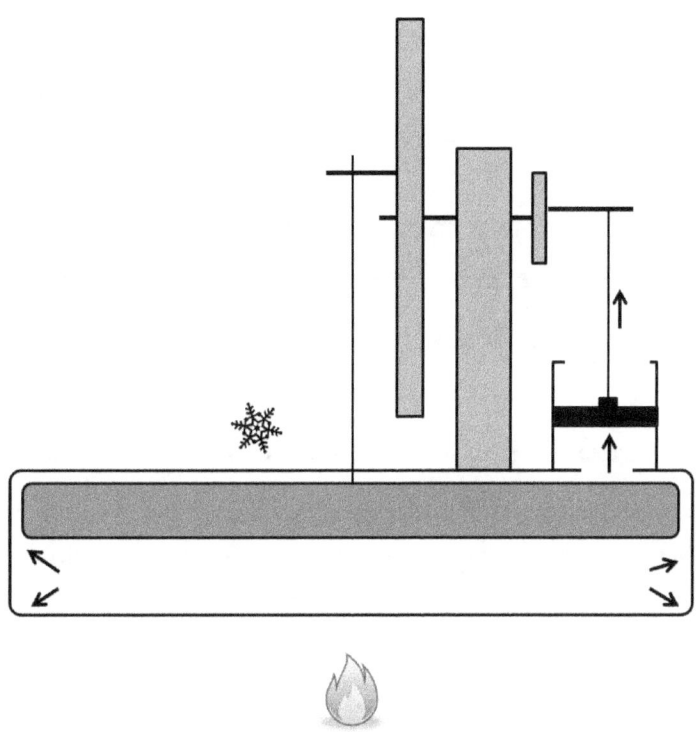

Warming Phase – The air is warming and expanding when the displacer is in the top of the pressure chamber.

When the engine is in the warming phase, the air is in the warm side of the engine. This causes the air to expand, which pushes the drive mechanism upward in an expansion power stroke.

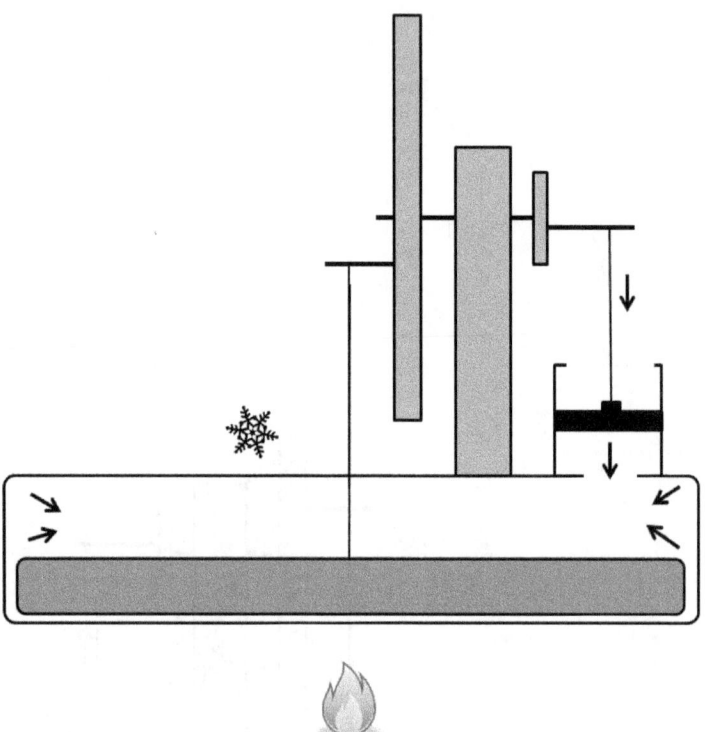

Cooling Phase - The air is cooling and contracting when the displacer is in the lower part of the pressure chamber.

When the engine is in the cooling phase, the air is on the cool side of the pressure chamber. The air contracts as it cools and this decreases the pressure inside the engine. This pulls down on the drive mechanism causing a contraction power stroke. The repeated pushing and pulling of the drive mechanism moves the crankshaft, and the stored energy in the moving flywheel carries the engine through the brief moments when there is no power stroke.

Stirling Engine Zen

I have now been helping people build Stirling Engines for several years. As people contact me and show me their work, I see a wide variety of skill levels demonstrated. While some people send in videos of running motors that look exactly like the ones in the books, others send me pictures of parts that somewhat resemble some of the parts described in the plans. These people are asking for help because their engines are not looking much like the ones in the book, and they don't work.

As a writer and a trainer, I want to constantly improve my ability to give directions so that more and more people experience success as they use my instructional material to build their own Stirling engines. Every failure that I get to help with is an opportunity for me to improve my instructions so that they make sense to everyone.

I have observed successful builders and I have observed those who experienced challenges. There are several notable differences between the two. Everyone brings a unique set of experiences, skills, and abilities to their Stirling engine project. People who have experienced previous success while working with these types of materials are likely to experience success with these projects. People who are new to working with sheet metal, acrylic, pliers, and music wire can also experience success, but will have more to learn along the way.

I have concluded that success with a project, such as the ones in this book, has as much to do with the *mindset* of the builder as it does with their previous experience. This state of mindfulness that helps people successfully complete these complex projects is what I call "Stirling Engine Zen."

What is Stirling Engine Zen? It is a combination of things related to how we choose to approach complex tasks.

Observe how each part contributes to the whole. Study the plans and instructions for each part before you build it. Create an understanding for yourself about how this part is supposed to function within the finished engine. What are the characteristics that make this part successful? If the part must be cut to precise measurement, or be made of a specific material, focus on those qualities and strive to meet the standard specified in the plans. If you elect to create your own alternative part or process, carefully examine how your changes to the original design could affect the performance of the engine.

Be patient. Take as much time as is needed for successful completion of each construction step. Be willing to make a complicated part more than once in your efforts to get it correct. Complicated tasks are much easier to complete the second or third time you do them. If you are making a complex part, such as a crankshaft, plan to make several of them and then use the best one in your engine. Avoid the temptation to use a poorly made part simply because you want to hurry on to the next step.

Value precision over approximation. Approach this project as if you were constructing a mechanical clock, working with as much precision as your tools will allow. While these projects do not require the precision of a watchmaker, precision works much better than approximation. Do your best to accurately reproduce the parts as illustrated in the book. The old saying, "If at first you don't succeed, get a bigger hammer" is not compatible with Stirling Engine Zen.

I think you will find that doing these three things will be very helpful as you strive to complete the projects in this book.

Skills Needed for These Projects

These instructions have been written with the goal of enabling many people with different skill sets and different tool sets to find success in building these models. You will not need to own a machine shop or have a lot of previous experience to complete a project from this book. There are some construction steps that may require minor adaptation to fit the tools and materials that are available. Many of those adaptations are described as options in the text. Skills and techniques will be described in detail so that the builder has enough information to complete the required tasks, even if this is the first time they have performed that particular task.

To be successful building one of these Stirling engines the builder must be able to perform the following tasks:

- Locate and obtain the needed parts and materials.
- Cut sheet metal, or find someone who can cut it for them.
- Cut clear acrylic sheet for engine sides and for the flywheel.
- Cut small pieces of wood.
- Drill holes accurately.
- Cut and bend wire.
- Measure accurately.
- Read, understand, and follow directions well.

Several assumptions are made about the potential builder of these engines. Several of these statements are probably true for someone building an engine from this book:

- They want to build a gadget that goes, like a Stirling engine.
- They are interested in Stirling engines or in science in general.
- They have access to basic tools for cutting and shaping the parts for these engines.
- They enjoy the challenge of figuring things out in the mechanical world.
- They may have built other types of Stirling engines before, and now they want to build an LTD engine.

Tools and Supplies

The list of tools needed for these projects is somewhat flexible. All of the designs in this book can be built with simple hand tools. The metal parts can be cut with tin snips, and other parts can be cut with hand saws and other small tools. Access to a drill press and a band saw would be useful, but those tools are not required.

Not every builder will have access to the same types of tools. The required tool set has been kept small intentionally so that these projects can be accomplished by many different people.

(On a personal note, I do have a motto that says, "Any job that requires you to buy a new tool is a job worth doing.")

You will need tools (or friends with tools) that can accomplish these tasks:

Cutting Sheet Metal

Aluminum or copper sheet metal parts will need to be cut into squares (or circles) to form the top plate and bottom plate of each engine. Something as simple as tin snips (or as fancy as a band saw) will work for this.

Cutting Acrylic Sheet

The engine sidewalls and flywheels are fashioned from clear acrylic. Acrylic is cut and drilled with the same tools used to cut and drill wood.

Drilling Holes

Small holes must be drilled in metal, wood, and plastic. Some of these holes must be drilled accurately at 90^0. This can be done with a hand drill, although a drill press would be the preferred choice if it is available. In some cases the holes to be drilled will be smaller than 1/16" (1.59 mm) which is the most common small drill bit in the US. (Remember that slogan about any job that makes you buy a new tool…)

Wood Working

Some of these engines use small wood parts. The wood must be cut and drilled.

Wire Cutting and Bending Tools

The music wire used in these models is best cut by scoring it with a file and then bending it to break it. Bending some small parts will require the use of pliers.

Adhesives

There are four adhesives that are recommended for these projects: Cyanoacrylate (Super Glue®), high temperature epoxy, clear epoxy, and silicone adhesive.

Super Glue® comes formulated in several thicknesses and cure times. A gel style of Super Glue® is very forgiving and will fill small gaps that thinner glues cannot cover.

There are several makers of high temperature epoxy. JB Weld and JB Kwik are well known as high temperature epoxy adhesives. Many brands will show the temperature rating on the package. Look for a product that can tolerate temperatures up to 400^0 F (204^0 C).

Clear five minute epoxy is a nice adhesive to use for clear parts that will not be exposed to high heat. The five minute set time makes it possible to complete multiple tasks in a single evening. People who are new to using epoxy may prefer a product that has a set time of 10 to 15 minutes. The extra time means you don't have to hurry quite so much for some gluing operations.

There are also many varieties of silicone adhesive. It works well for bonding uneven surfaces or for bonding joints where flexibility is desired. Marine supply stores offer a variety of silicone sealants and adhesives. The package will often rate and compare the holding power for each formulation. Some formulations are almost permanent after they cure, and others are more easily removed.

Auto supply stores sell a variety of silicone sealants that can tolerate extreme heat. They are used for making or sealing engine gaskets. They hold parts together moderately well but are not intended for permanent installation. Either marine or automotive silicone will work. The automotive silicone will be more affordable.

Measuring Tools

A simple ruler (metric or inches) is the minimum requirement. A micrometer is handy at times, but is not required.

Note: Metric equivalents are based on the mathematical equivalent of the measurement in inches and may not reflect the dimensions of actual products in the market in European countries. For instance, the metric equivalent of 1/4" acrylic sheet is displayed at 6.35 mm, while in reality the European reader will be looking for something close to a 6 mm acrylic sheet.

Parts Needed for These Projects
Teflon tube (PTFE) suppliers   http://www.portplastics.com http://www.zeusinc.com

- 0.0625" (1.6mm) Aluminum Sheet (top and bottom plates)
- Styrofoam sheet insulation (displacer)
- Black paint (Engine body)
- 1/4" (6 mm) Acrylic Sheet (side wall)
- 0.06 Acrylic Sheet (1/16") (flywheel)
- 1/16" Shaft Collar
- 3/4" PVC pipe (drive cylinder)
- 0.015" Music Wire (pushrods)
- 0.0625" (1/16") (1.6 mm) Music Wire (axle)
- Latex Rubber Glove (drive diaphragm)
- Duct Tape (connecting rod joint)
- Super Glue®
- Clear 5-minute epoxy
- Black JB Quick high temperature epoxy
- RTV Silicone automotive adhesive
- Wood (for the pedestal)
- AWG-14, 0.066" (1.6764 mm)heavy wall extruded Teflon tube (bushings)
- #24 (0.022" ID) (0.5588 mm) Teflon tube (gland)

Finding the Parts

0.0625" (1.6 mm) Aluminum Sheet (Top and Bottom Plates)
Aluminum sheet metal can be purchased at hardware stores, model shops, salvage yards, or from online retailers such as Amazon.com. (Copper is an excellent thermal conductor and may be substituted for the aluminum.)

Styrofoam Sheet Insulation (Displacer)

The Styrofoam used for building the models pictured in this book was cut from a sheet of foam insulation purchased at a local building supply store. It is available in a variety of thicknesses. The foam was cut to shape with a hot wire foam cutter described near the end of this book.

Styrofoam is often used as packing material. It may be possible to salvage Styrofoam from packing materials with the use of a hot wire foam cutter.

Styrofoam can have a relatively low melting point. This is usually not a problem with low temperature differential engines. Choose a material for the displacer that will be compatible with the temperatures that will be used to run the motor. White Styrofoam is very light in weight, which is good in this application. But white Styrofoam will melt at about 250^0 F (120^0 C). An engine with a displacer made of this material should never be run on a surface that is hotter than boiling water.

There are different kinds of Styrofoam that have a higher melting point, and there are other materials that can be used to make the displacer. Alternative displacer materials are discussed elsewhere in the book. The white Styrofoam insulation was chosen this time because of its light weight and because the engine will not be exposed to temperatures over 212^0 F (100^0 C).

Black Paint (Engine Body)

The top and bottom plates of the pressure chamber can be painted with flat black spray enamel. This can be found at paint stores, hardware stores, or building supply stores. Painting the surfaces black is optional and will improve performance when using sunlight to power the engines.

1/4"(6.35 mm) Acrylic Sheet (Side Wall)

Acrylic sheet is frequently used as an alternative to glass for window glazing. Stores that sell acrylic window glass will sometimes also carry the thicker material, such as the 1/4" (6.35 mm) material recommended here. Acrylic in this thickness works better than the thinner acrylic for building the sidewalls of the engine. It is possible to substitute other thicknesses of acrylic if this exact material cannot be found. Metric thicknesses of 6 mm or greater can be used.

It may be necessary to purchase this material from a specialty supplier. Most big cities have a shop or two that specializes in plastics. One retailer in the US is Tap Plastics (http://www.tapplastics.com/). Acrylic sheet can also be purchased through Amazon.com.

0.06 Acrylic Sheet (1/16", 1.59 mm) (Flywheel)

Whereas the pressure chamber side walls are made from material that is thicker than window glazing, the flywheel is made from acrylic that is slightly thinner than window glazing. Like the thicker sheet, it is available from specialty plastic stores and from online retailers. This material is frequently used with picture frames and can often be found at stores where picture frames are made or sold.

1/16" (1.59 mm) Shaft Collar

These small shaft collars were found in a shop that specializes in radio controlled models and model parts. The package is labeled, "1/16" Nickel-Plated Brass Wheel Collars." These shaft collars were selected for these projects because they simplify the process of attaching things to the 1/16" (1.59 mm) shaft used to make the engine axles.

3/4" PVC Pipe (Drive Cylinder)

The PVC pipe was purchased at a building supply store. PVC pipe is available in the USA in a variety of thicknesses, and this one is the lighter thin-walled version. It is labeled and sold as 3/4" pipe, but the actual outside diameter of the pipe is 1 1/16" (27 mm). The inside diameter is just over 7/8" (22.23 mm). Other types of tubing of a similar diameter may be used as a substitute if this exact type of plastic tube (pipe) cannot be found.

0.015" Music Wire (Pushrods), and 0.0625 (1/16") Music Wire (Axle)

Music wire can be purchased at model shops and from well stocked hardware stores. It is also available online. One reputable dealer that seems to carry a full inventory of music wire is Tower Hobbies (http://www.towerhobbies.com/). A guitar string can be used as a substitute for the 0.015" music wire.

Latex Rubber Glove (Drive Diaphragm)

The rubber gloves used to make these engines were purchased in the paint department at the local building supply store. Rubber gloves are also available at medical supply stores and drug stores.

Duct Tape (Connecting Rod Joint)

This is a vinyl cloth tape that is flexible and very sticky. It was originally designed for holding duct work together, but it has become a common tape for repairs of any kind that need strong heavy tape. This is available in hardware and building supply stores. Any flexible cloth tape with good adhesion will work.

Clear 5-Minute Epoxy

Two-part epoxy glue can be purchased in hardware stores, building supply stores, office supply stores, and sometimes even in the housewares department of the local grocery store. There are noticeable differences in the quality of the various brands. Model shops will often sell the better brands. Avoid the cheap generic epoxy as it will not bond well and will melt when exposed to heat.

Black JB Quick® High Temperature Epoxy

This is another two-part epoxy glue that cures quickly. It handles the heat very well and has exceptional bonding strength. "JB Quick®" is a brand name. There are other brands of high temperature epoxy glue that will also work well.

RTV Silicone Automotive Adhesive

Auto supply stores will often carry a variety of different formulations of RTV Silicone. Read the labels and find one that has good holding power and low fumes. Some formulas have nasty fumes that make them hard to work with in confined areas. Silicone is used for sealing the pressure chamber for several reasons. It allows motion as some parts may expand with heat, it tolerates heat well, and it is easy to remove if the seal ever needs to be opened for repairs. There are other sources for silicone adhesive that will work equally well, including marine supply stores, and the paint department at the building supply store.

Wood (Pedestal)
The engines with wooden pedestals use a piece of wood that is approximately 3/4" (19.05 mm) square by 3" (76.2 mm) long. This can be cut from a common 3/4" (19.05) (or similar thickness) board.

AWG-14 0.066" Heavy Wall Extruded Teflon Tube (Bushings), and #24 (0.022" ID) Teflon Tube (Gland)
Teflon tube is used to make the bushings for the axle bearings, and it is also used to create the gland opening in the pressure chamber. Teflon tube was chosen for these parts because it provided the best results in testing. The testing process and the test results are detailed in another chapter of this book. There are other materials that can be substituted if the Teflon cannot be located. Read the test results and pick a material that performs well if you must make a substitution.

The Teflon tube for these projects was acquired from two different sources. The AWG-14 Heavy Wall Extruded Tube was found online at the manufacturer's website, http://www.zeusinc.com/. Zeus manufactures and sells precision plastic tubing in many sizes and formulations. The Zeus company does not sell small quantities of their product, but they will help customers find a dealer who can sell small quantities if you contact them through their website. Zeus has operations in the United States and in Europe.

The smaller #24 Teflon tube was ordered online at http://www.ebay.com/.

It is very important that the Teflon tube sizes are matched to the sizes of the axle and the displacer pushrod. The AWG14 Teflon tube has an inside diameter of 0.066." The axle diameter is 0.0625", leaving a clearance of 0.0035" (0.0889 mm). This allows the axle to turn freely inside the Teflon bushing while still maintaining a good fit.

The #24 Teflon tube has an inside diameter of 0.022." The music wire used for the pushrod has a diameter of 0.015", leaving a clearance of 0.007" (0.1778 mm).

These clearances are sufficient to make a smooth running Stirling engine. The engine can be built with smaller clearances if care is taken to ensure that there is no increase in binding or friction.

Alternative construction methods are outlined near the end of the book. You will find alternative approaches for bearings, drive cylinders, displacers, and more.

Construction Sequencing
In some cases it is quite possible to alter the sequence of the construction steps to meet the needs of the builder. Each model is constructed by creating the baseline parts to which the other parts must be fit. There will be some variance in dimensions because of the nature of any hand built project. The key to good sequencing is knowing which parts can be easily altered to accommodate for parts that don't measure up exactly to the plan.

There are a few parts on some motors that must be built to an exact specification. These more demanding parts are usually constructed early in the process, and other parts are designed to fit within the space requirements of the more demanding parts.

For instance, all of the engines have a displacer pushrod that must have a gland in the center of the top plate. The smaller engines have very little space for positioning the pedestal, crankshaft, flywheel, and drive cylinder. With the drive cylinder placed at the outside edge of the top plate, the remaining parts must be made to fit and function well in the remaining space. It is important to know how much space is really available before building these parts. In this case it is necessary to determine the position and placement of the drive cylinder before constructing the pedestal and flywheel mechanism that must fit between the drive cylinder and the displacer pushrod.

The larger engines have more surface area available and provide a bit more flexibility for placement of the drive mechanism. For these engines the drive mechanism can be assembled before the drive cylinder is located. The drive cylinder can then be positioned so that it aligns perfectly with the drive mechanism.

Dimensions
It is recommended that these models be built as close to the listed dimensions as is possible. There will be times, however, when a dimension may need to be altered by a small amount to accommodate for a material choice or human error. If a different thickness of acrylic sheet is used to make the engine sidewall than what is in the plans, the displacer dimensions must be adjusted to accommodate. If the diameter of the round sidewall engine is not exactly the same dimension as the plans, the diameter of the displacer must be adjusted to match. These circumstances will be discussed in the instructions at those points where adjustment is likely to be needed.

Design Note
The common theme between these designs is the central pedestal and the straight axle. This drive mechanism is fairly easy to assemble. This simplifies the construction process while also increasing the chances for building a successful engine.

The plans show how to build this design in large, small, square, and round configurations. The round engines have a similar look and feel as the designs traditionally built using a metal lathe. Choose one of these designs if you desire the traditional aesthetics of a round engine. The square designs will have more surface area for heating and cooling, and may be a little more efficient as a result. The square engine body is simpler to build and goes together a bit faster.

Engine #4: Small Square Engine 4" (10.2 cm)

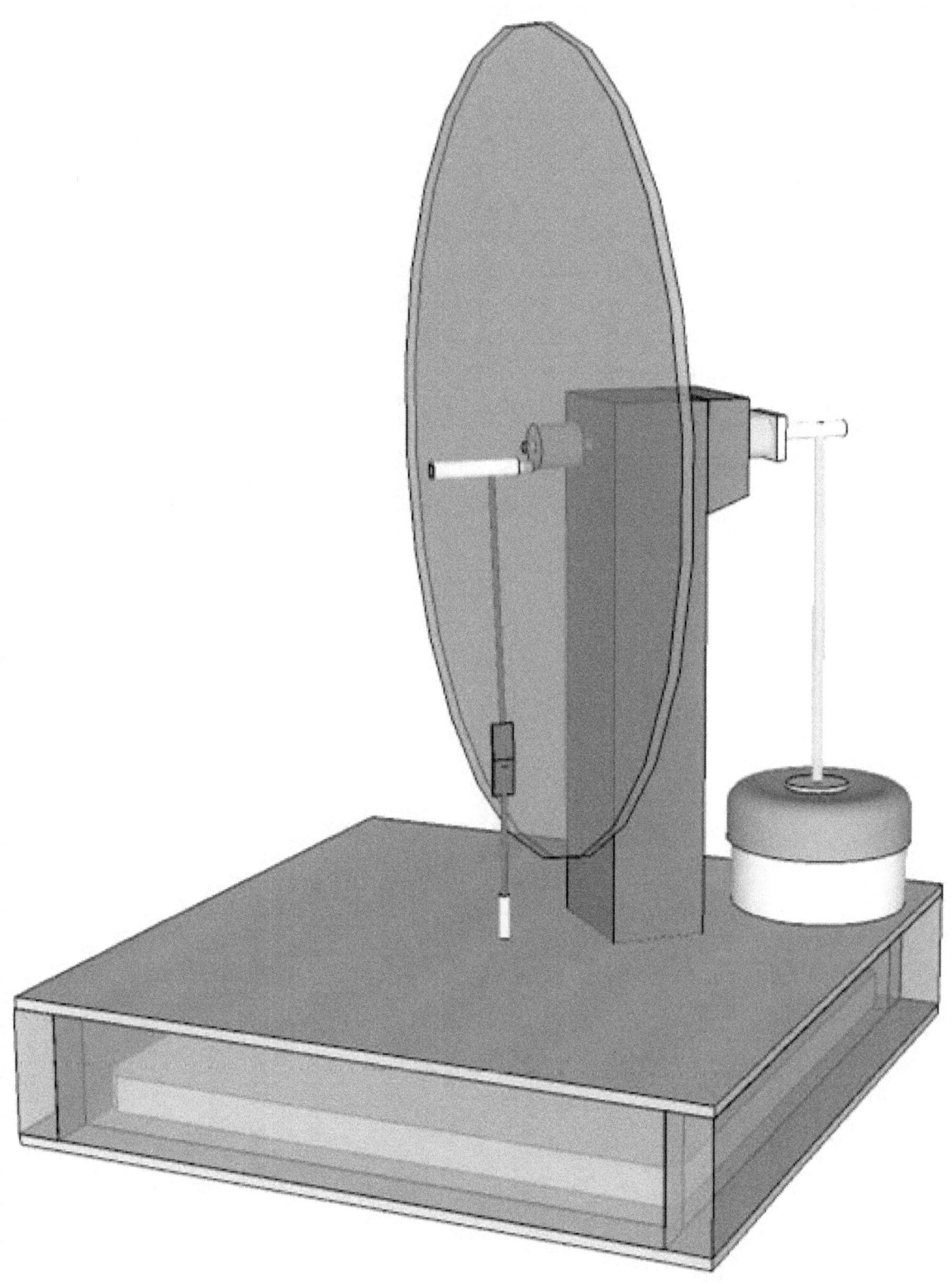

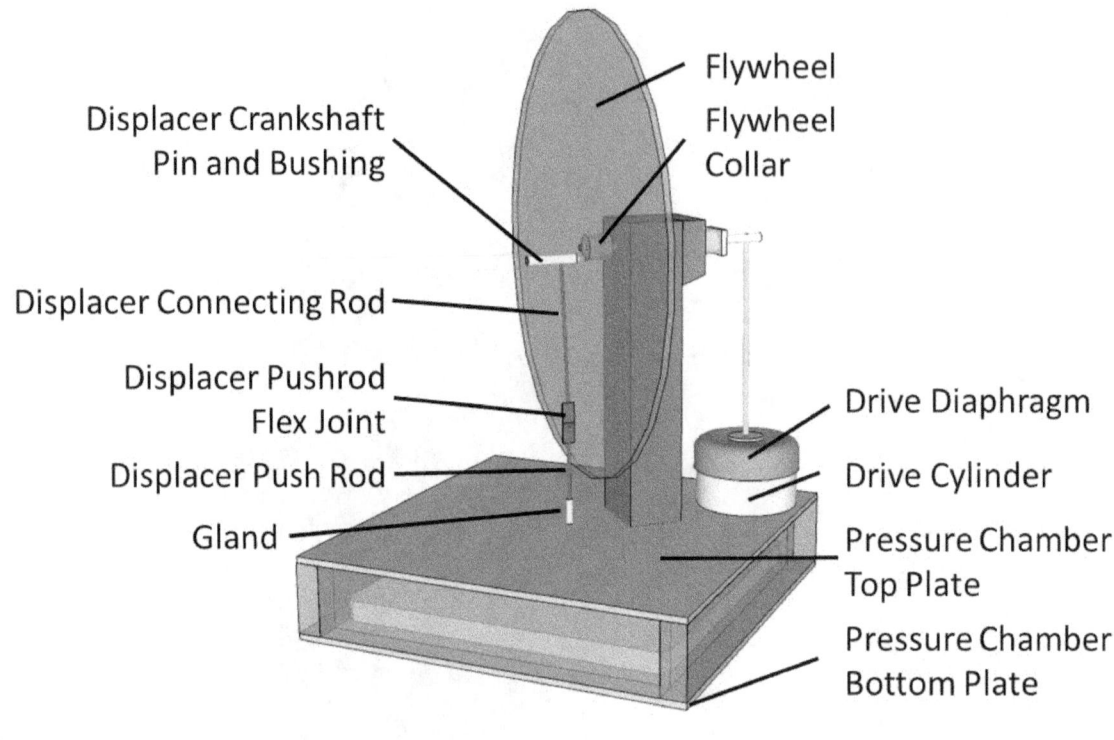

Flywheel

Displacer Crankshaft Pin and Bushing

Flywheel Collar

Displacer Connecting Rod

Displacer Pushrod Flex Joint

Drive Diaphragm

Displacer Push Rod

Drive Cylinder

Gland

Pressure Chamber Top Plate

Pressure Chamber Bottom Plate

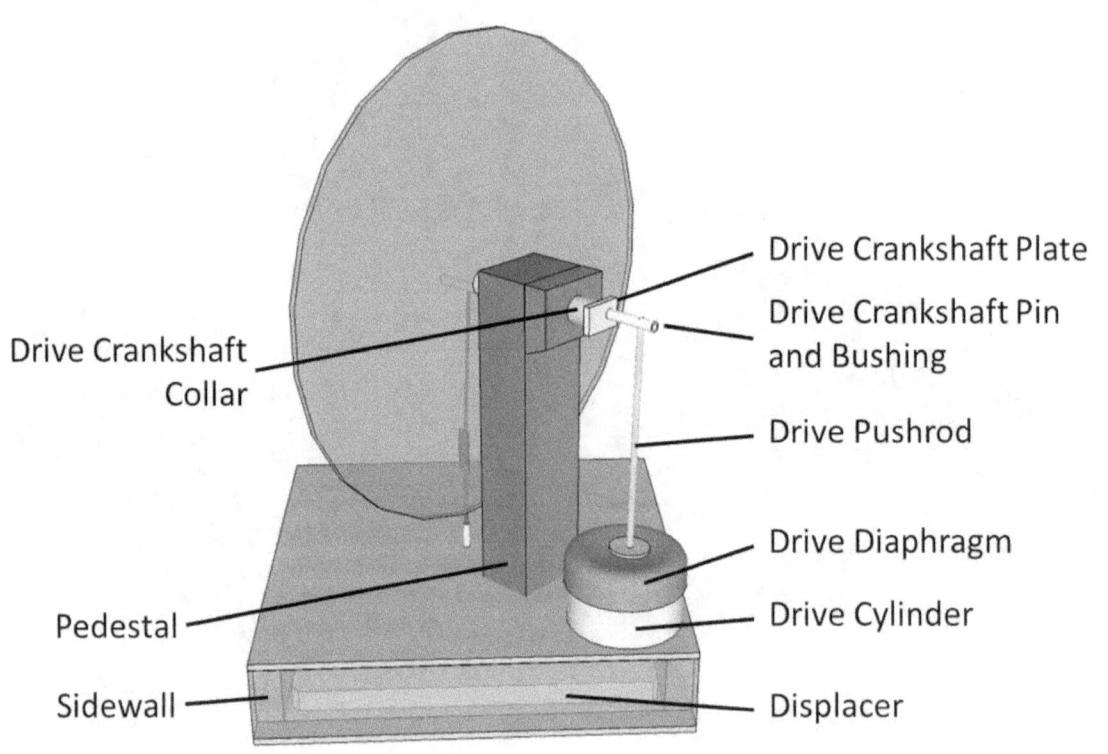

Drive Crankshaft Plate

Drive Crankshaft Pin and Bushing

Drive Crankshaft Collar

Drive Pushrod

Pedestal

Drive Diaphragm

Drive Cylinder

Sidewall

Displacer

Parts and Materials

Part #	Part Name	Description
01	Flywheel	Clear Acrylic Sheet, round, 0.06" x 4 1/2" (11.43 cm) diameter*
02	Flywheel Collar	Shaft Collar, 1/16" (1.59 mm)
03	Pressure Chamber Bottom Plate	Aluminum Sheet, 0.062" x 4" (10.16 cm) x 4" (10.16 cm)
04	Pressure Chamber Top Plate	Aluminum Sheet, 0.062" x 4" (10.16 cm) x 4" (10.16 cm)
05	Displacer	Styrofoam, 0.20" (5 mm) x 3 3/8" (8.57 cm) x 3 3/8" (8.57 cm)
06	Displacer Crankshaft Pin	Music Wire, 0.0625 x 1/2" (12.7 mm)
07	Displacer Crankshaft Bushing	AWG-14 (0.066") Heavy Wall Extruded Teflon Tube, 3/8" (9.53 mm)
08	Long Sidewall (2 pieces)	Clear Acrylic, 1/4" (6.35 mm)** x 11/16" (17.46 mm) x 4" (10.16 cm)
09	Short Sidewall (2 pieces)	Clear Acrylic, 1/4" (6.35 mm)** x 11/16" (17.46 mm) x 3 1/2" (8.89 cm)
10	Pedestal	Wood, 3/4" (19.05 mm) x 5/8" (15.88 mm) x 3" (7.62 cm), and 3/4" (19.05 mm) x 3/4" (19.05 mm) x 1/4" (6.35 mm)
11	Displacer Connecting Rod	Music Wire, 0.015" x 3 1/2" (8.89 cm)
12	Displacer Pushrod	Music Wire, 0.015" x 6" (15.24 cm)
13	Displacer Pushrod Flex Joint	Duct Tape, 1/8" (3.18 mm) x 1/2" (12.7 mm), 2 pieces
14	Drive Crankshaft Pin	Music Wire, 0.0625 x 1/2" (12.7 mm)
15	Drive Crankshaft Bushing	AWG-14 (0.066") Heavy Wall Extruded Teflon Tube, 3/8" (9.53 mm)
16	Drive Crankshaft Collar	Shaft Collar, 1/16" (1.59 mm)
17	Drive Crankshaft Plate	Aluminum Sheet, 0.062" 1/4" (6.35 mm) x 3/8" (9.53 mm)
18	Gland	Teflon Tube, #24 (0.022" ID) x 7/16" (11.11 mm)
19	Drive Cylinder	PVC Pipe, 1" (25.4 mm) ID x 5/8" (15.88 mm)
20	Drive Diaphragm	Latex glove fingertip
21	Drive Pushrod	Music Wire, 0.015" x 3 1/2" (8.89 cm)
22	Axle	Music Wire, 0.0625"x 1 1/2" (38.1 mm)
23	Axle Bushings (2 pieces)	AWG-14 (0.066") Heavy Wall Extruded Teflon Tube, 3/16" (4.76 mm)

*Flywheel thickness can vary slightly from the prescribed thickness of 0.06", depending upon the material available to the builder. Acrylic sheet commonly sold in the US market for window glazing is slightly thicker (0.093") and is used for the flywheels of some of the engines in this book. Thin polystyrene sheet material used in covering framed pictures is only 0.05" thick and will suffice, but can be difficult to work with.

** It is also possible to make the sidewalls from slightly thinner 3/16" (4.76 mm) acrylic sheet.

Drawings and Dimensions

Note: Images are not drawn to scale.

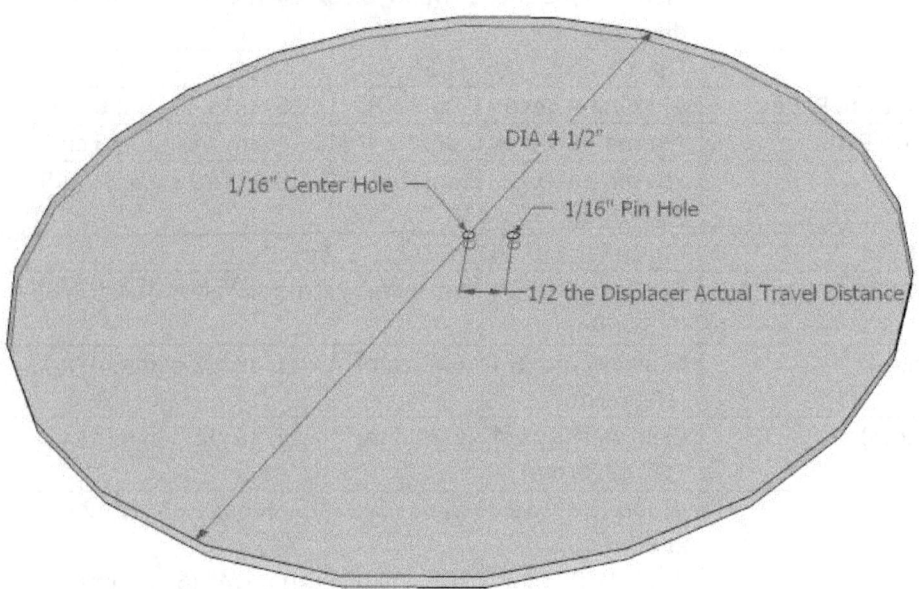

Part # 01 Flywheel - Clear Acrylic Sheet, round, 0.06" x 4 1/2" (11.43 cm) diameter

Part # 02 Flywheel Collar- 1/16" (1.59 mm) Shaft Collar
Part # 16 Drive Crankshaft Collar - 1/16" (1.59 mm) Shaft Collar

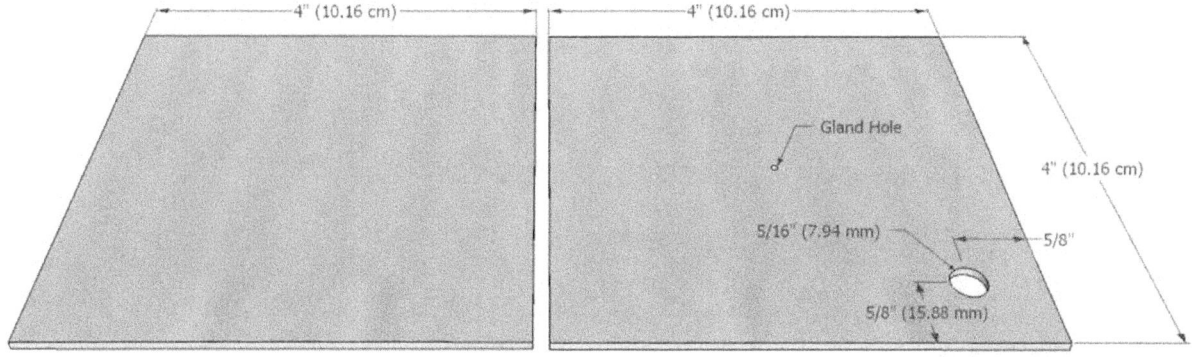

Part # 03 Pressure Chamber Bottom Plate   **Part # 04 Pressure Chamber Top Plate**
Aluminum Sheet, 0.062" x 4" (10.16 cm) x 4" (10.16 cm) (2 pieces)

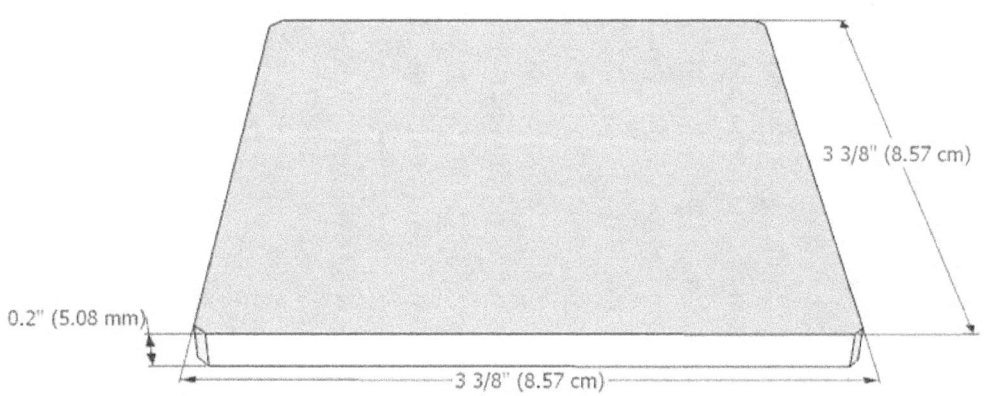

Part # 05 Displacer - Styrofoam, 0.20" x 3 3/8" (8.57 cm) x 3 3/8" (8.57 cm)

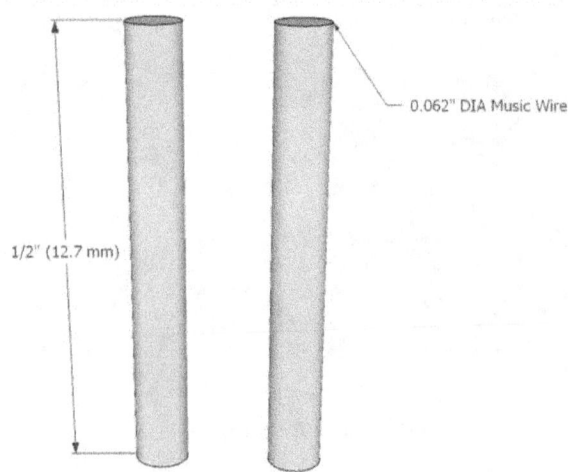

Part # 06 Displacer Crankshaft Pin - Music Wire, 0.0625 x 1/2" (12.7 mm)
Part # 14 Drive Crankshaft Pin - Music Wire, 0.0625 x 1/2" (12.7 mm)

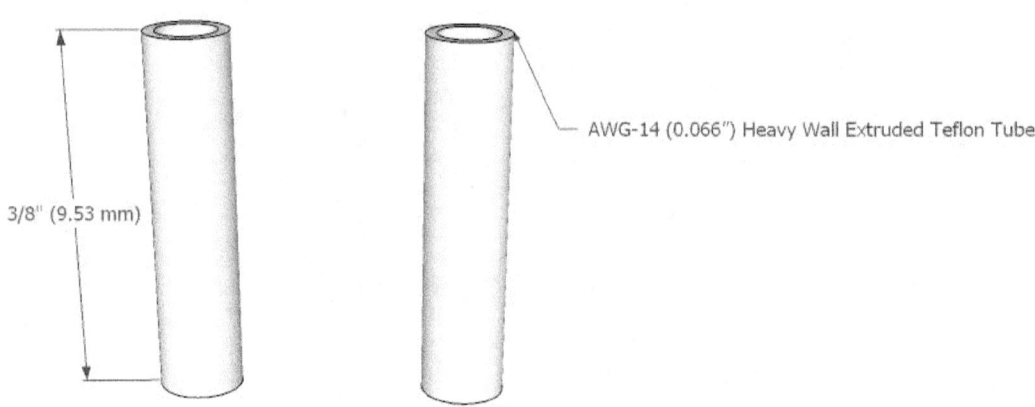

Part # 07 Displacer Crankshaft Bushing - AWG-14 (0.066") Heavy Wall Extruded Teflon Tube, 3/8" (9.53 mm)
Part # 15 Drive Crankshaft Bushing - AWG-14 (0.066") Heavy Wall Extruded Teflon Tube, 3/8" (9.53 mm)

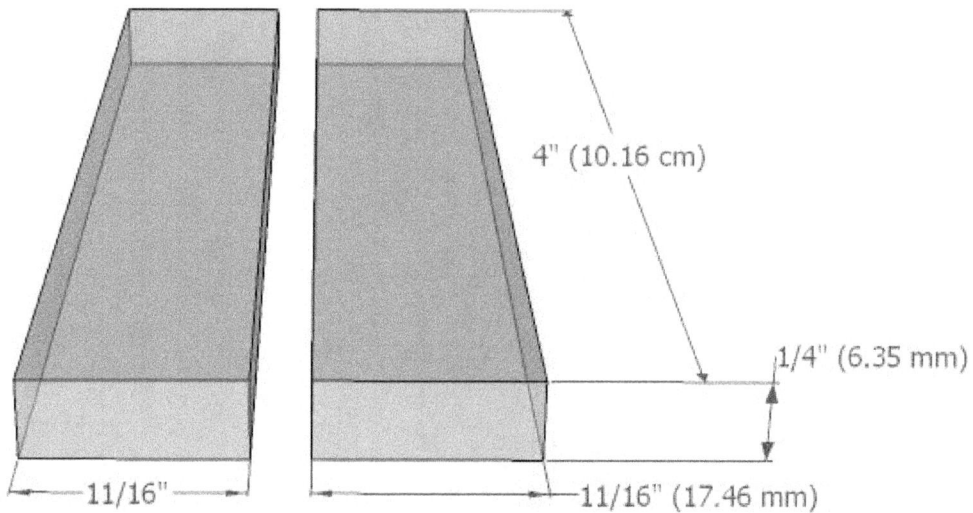

Part # 08 Long Sidewall - Clear Acrylic, 1/4" (6.35 mm) x 11/16" (17.46 mm) x 4" (10.16 cm) (2 pieces)

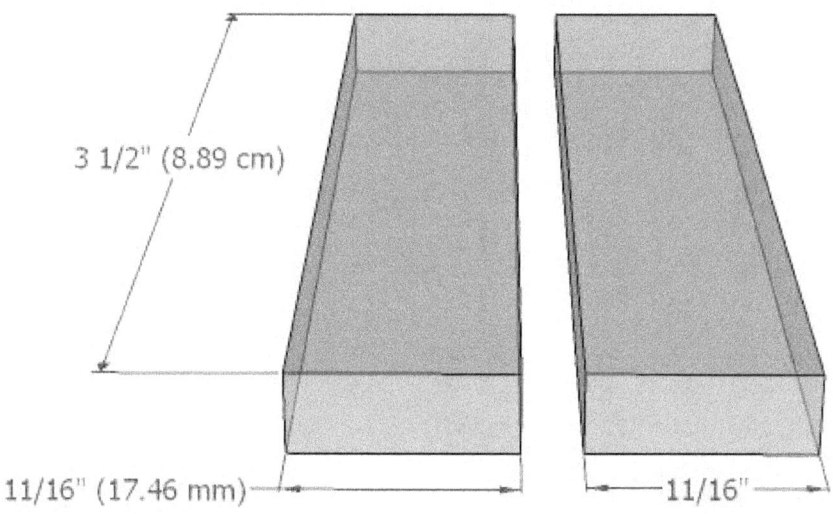

Part # 09 Short Sidewall - Clear Acrylic, 1/4" (6.35 mm)** x 11/16" (17.46 mm) x 3 1/2" (8.89 cm) (2 pieces)

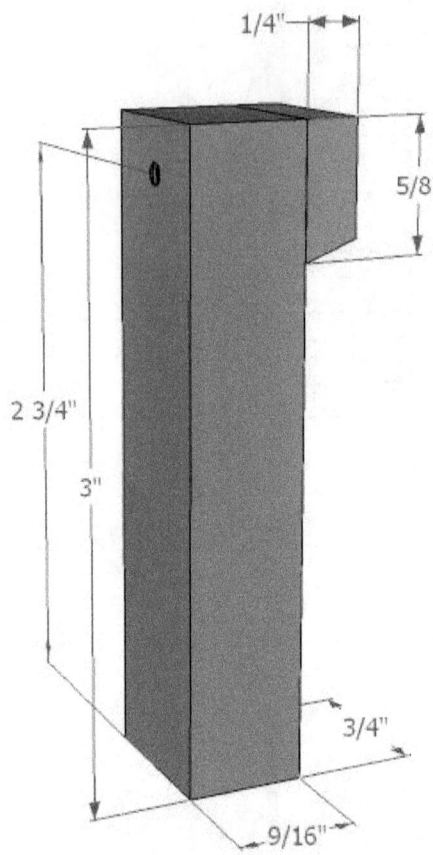

Part # 10 Pedestal - Wood, 3/4" (19.05 mm) x 5/8" (15.88 mm) x 3" (7.62 cm), and 3/4" (19.05 mm) x 3/4" (19.05 mm) x 1/4" (6.35 mm)

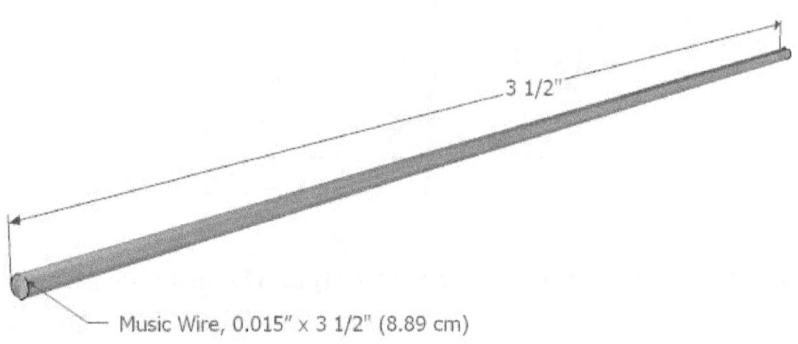

Music Wire, 0.015" x 3 1/2" (8.89 cm)

Part # 11 Displacer Connecting Rod - Music Wire, 0.015" x 3 1/2" (8.89 cm)
Part # 21 Drive Pushrod - Music Wire, 0.015" x 3 1/2" (8.89 cm)
(2 Pieces)

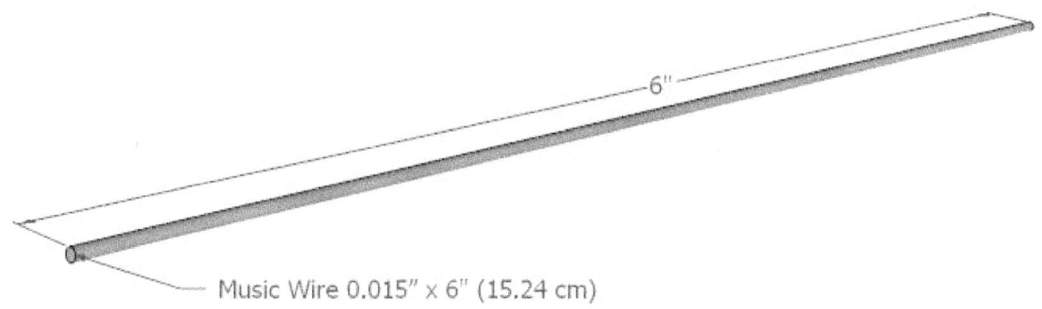

6"

Music Wire 0.015" x 6" (15.24 cm)

Part # 12 Displacer Pushrod - Music Wire 0.015" x 6" (15.24 cm)

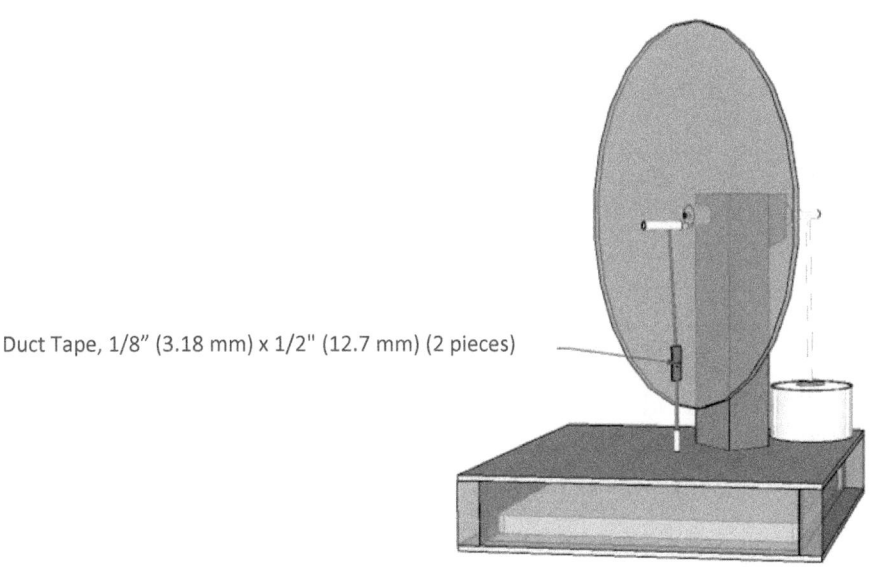

Duct Tape, 1/8" (3.18 mm) x 1/2" (12.7 mm) (2 pieces)

Part # 13 Displacer Pushrod Flex Joint - Duct Tape, 1/8" (3.18 mm) x 1/2" (12.7 mm) (2 pieces)

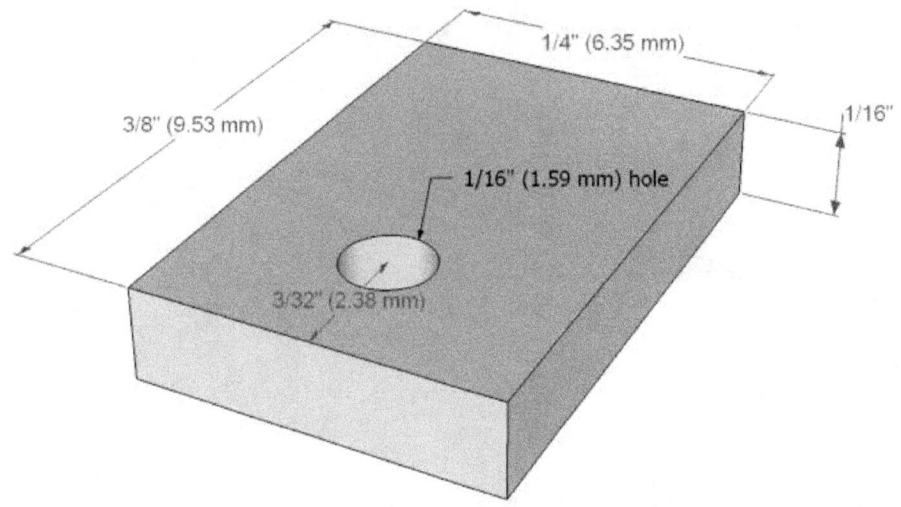

Part # 17 Drive Crankshaft Plate - Aluminum Sheet, 0.062" (1/16", 1.59 mm), 1/4" (6.35 mm) x 3/8" (9.53 mm)

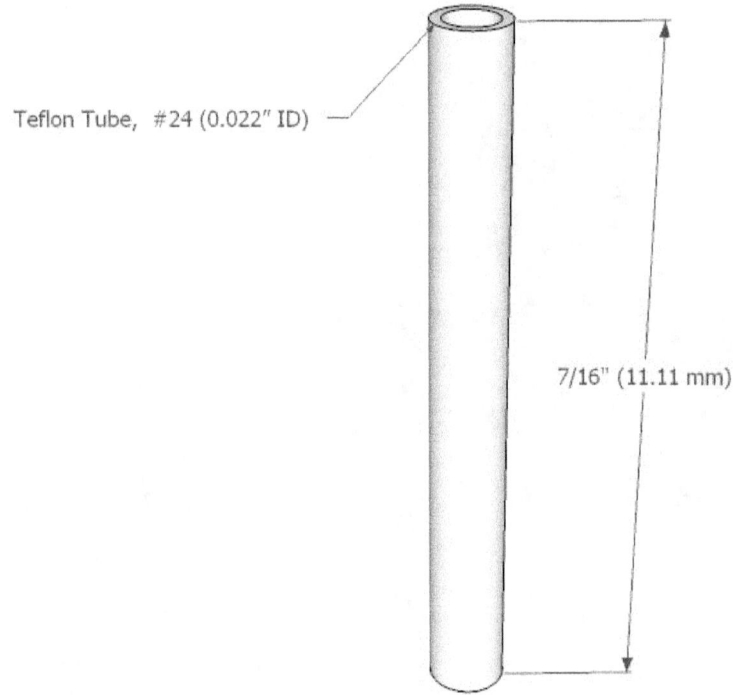

Part # 18 Gland - Teflon Tube, #24 (0.022" ID) x 7/16" (11.11 mm)

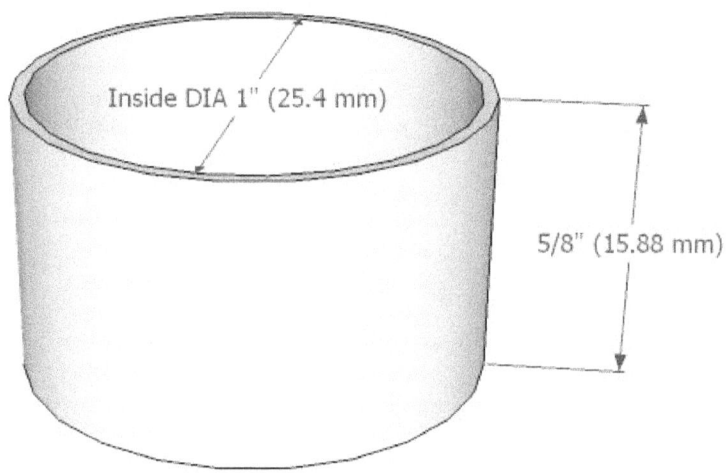

Inside DIA 1" (25.4 mm)

5/8" (15.88 mm)

Part # 19 Drive Cylinder - PVC Pipe, 1" (25.4 mm) ID x 5/8" (15.88 mm)

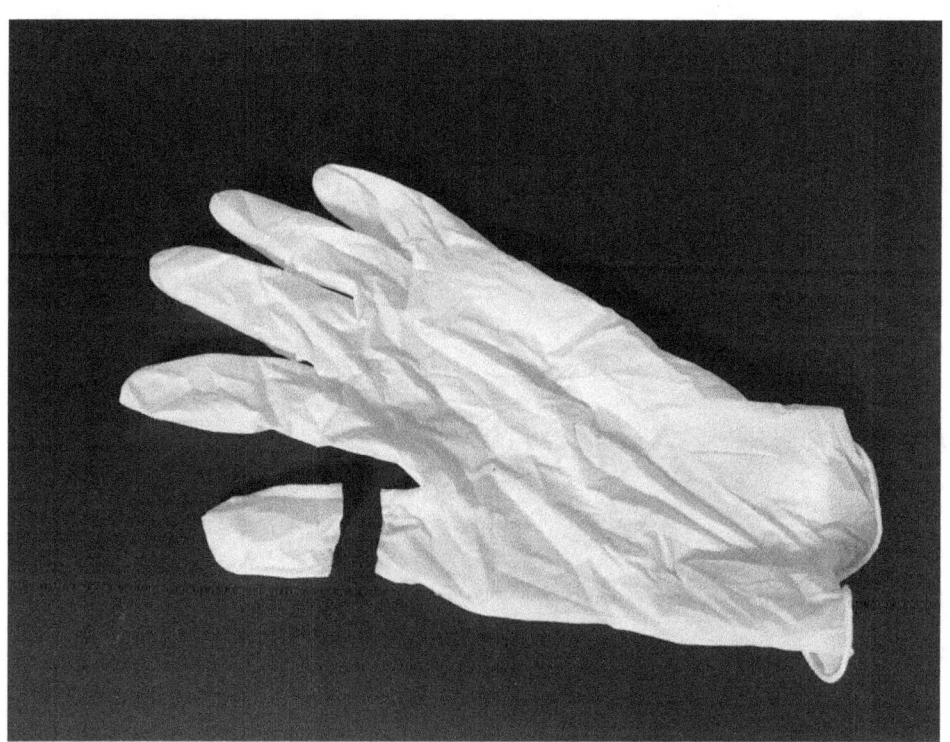

Part # 20 Drive Diaphragm - Latex glove fingertip

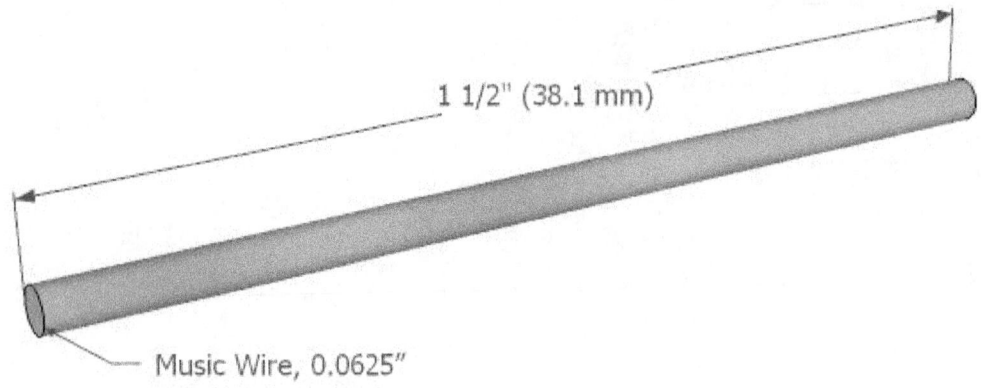

Music Wire, 0.0625"

Part # 22 Axle - Music Wire, 0.0625"x 1 1/2" (38.1 mm)

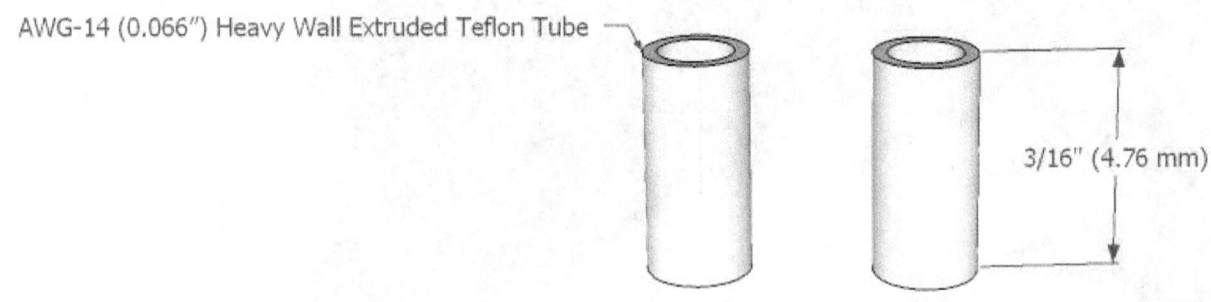

AWG-14 (0.066") Heavy Wall Extruded Teflon Tube

3/16" (4.76 mm)

Part # 23 Axle Bushings - AWG-14 (0.066") Heavy Wall Extruded Teflon Tube, 3/16" (4.76 mm) (2 pieces)

Assembly Instructions for Small Square Engine

Make the Pressure Chamber Top and Bottom Plates

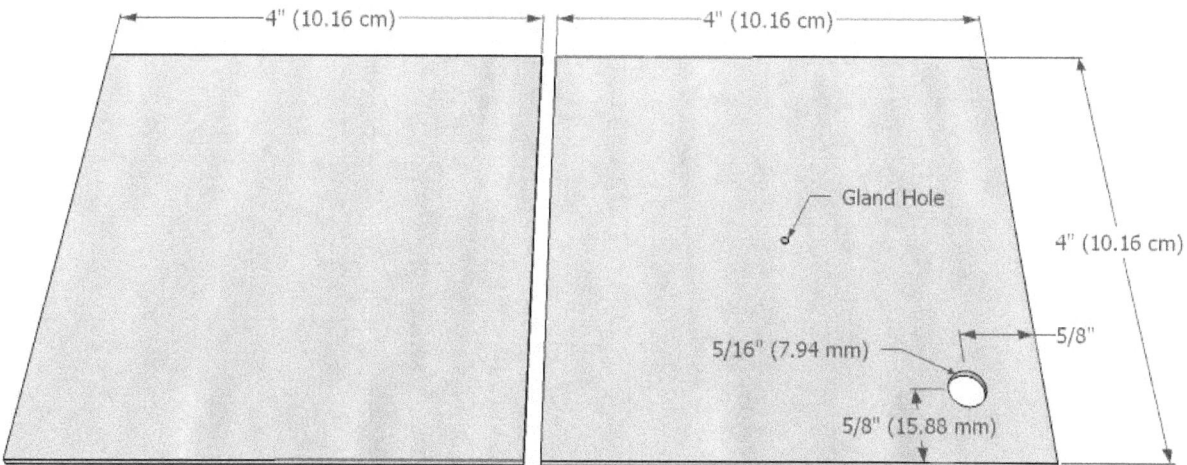

Cut two aluminum plates to form the top and bottom plates of the pressure chamber. Plate thickness is 0.0625" (1.59 mm). The plates are square, measuring 4" (10.16 cm) x 4" (10.16 cm).

Wash the metal parts with soap and water, rinse well and allow them to dry.

Paint all surfaces of the top and bottom plates (if desired). Use spray enamel that is suitable for painting metal.

Use a straight edge to find the exact center of the top plate. Use a center punch to mark the spot. Drill a hole that is the same diameter (or slightly larger) than the outside diameter of the Teflon tube that will be used for the displacer gland. The size will vary depending upon the material chosen.

Mark the position for the drive cylinder hole. The hole is located in one corner of the top plate, 5/8" (15.88 mm) from each side. The size of the hole is 5/16" (7.94 mm). Mark the hole position with a center punch and then carefully drill the hole.

Make the Pressure Chamber Side Walls
Cut two short sidewalls and two long sidewalls.

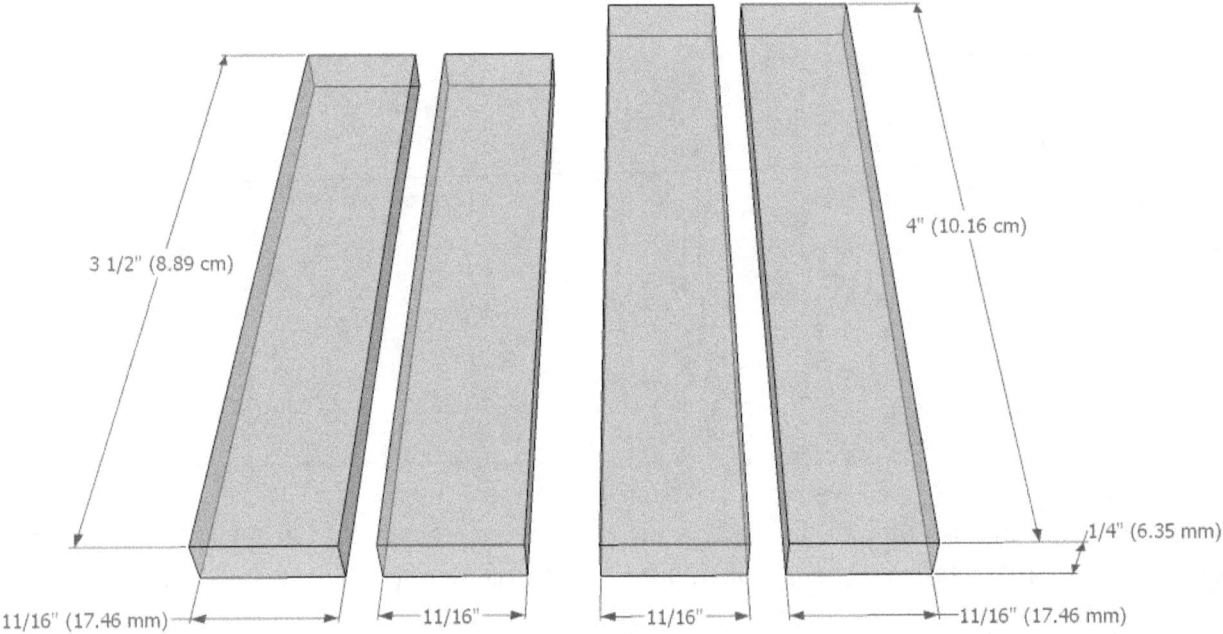

The sidewalls are cut from clear 1/4" (6.35 mm) acrylic sheet. The height of the sidewalls is 11/16" (17.46 mm). The length of the short sidewalls is 3 1/2" (8.89 cm). The length of the long sidewalls is 4" (10.16 cm).

It is also possible to make the sidewalls from slightly thinner 3/16" (4.76 mm) acrylic sheet.

Attach the Sidewalls to the Pressure Chamber Bottom Plate
Carefully check the fit of all the parts before assembly and use enough adhesive to ensure a good seal. Use clear five-minute epoxy to attach the acrylic sidewalls to the pressure chamber bottom plate. The sidewalls must be attached so that the pressure chamber is air-tight. The corner joints must also be sealed with glue. A small bead of glue oozing out of the joint indicates that the joint is being sealed well. Too much oozing glue, however, may interfere with the movement of the displacer inside the pressure chamber.

The top plate will _not_ be glued on at this time. The top plate may be used to help hold the sidewalls in place as the glue hardens.

Note: The glue will leave a permanent mark anywhere that it comes in contact with the acrylic. Be very careful not to create too many fingerprints on the acrylic during the gluing process, as these may be permanent.

Note: Silicone adhesive is another option for gluing the sidewalls to the pressure chamber. Silicone adhesives tolerate heat better than most clear epoxy glues, but they do not hold as well. Silicone is especially useful if there is a need to disassemble the engine for any reason.

The pressure chamber top plate, bottom plate, and sidewalls are ready for assembly.

Dry fit the parts before gluing and make any adjustment necessary to create a good fit.

Epoxy has been applied to attach the pressure chamber sidewalls to the bottom plate. The top plate is not being attached at this time. The top plate is holding the sidewall pieces in place as the glue cures.

Cut the Foam Displacer Panel

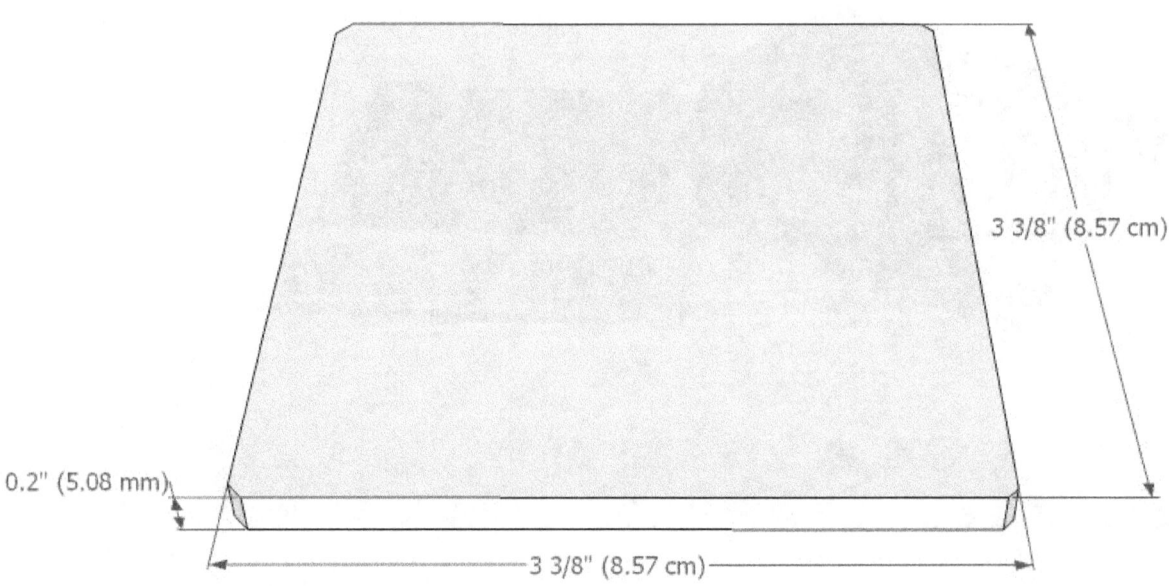

3 3/8" (8.57 cm)

0.2" (5.08 mm)

3 3/8" (8.57 cm)

The displacer is made from Styrofoam that is cut to a thickness of 0.20" (13/64", 5.08 mm). For the best results, cut the foam with a hot wire foam cutter similar to the one described elsewhere in this book. The corners of the foam displacer have been rounded to prevent contact with the inside of the pressure chamber.

A hot wire foam cutter uses Nichrome wire that is heated by a small electrical current to cut the foam. The framework of the hot wire foam cutter holds tension on the wire to keep it straight, and a small electrical transformer provides the current to heat the wire. Foam cutters can be purchased from craft suppliers, or one can be built using simple parts. Plans for a homemade hot wire foam cutter are provided later in this book.

The dimensions of the displacer provide clearance between the displacer and the sidewall of the pressure chamber. This clearance needs to be 1/16" (1.59 mm) to 1/8" (3.18 mm) between the side of the displacer panel and the pressure chamber sidewall. Measure the internal dimensions of the pressure chamber and make any necessary adjustments to the size of the displacer before cutting.

Attach the Displacer Pushrod
Cut a piece of 0.015" music wire to a length of approximately 6" (15.24 cm). Small diameter music wire can be cut with ordinary wire cutters.

Make a 90^0 bend 1/2" from one end of the music wire.

A simple jig is used to hold the wire at the correct angle while attaching it to the foam displacer. Make the jig by drilling a small hole through a piece of flat wood. Drilling the hole in the jig will require a small diameter drill bit and a drill press. Choose a bit size that will provide a snug fit for the music wire.

If small drill bits under 1/16" (1.59 mm) are not available, use a 1/16" (1.59 mm) drill bit and insert a piece of Teflon tube into the hole to make a better fit for the music wire.

Use a straight edge to find and mark the exact center of the displacer foam. Pierce the foam with the long straight end of the displacer pushrod and press it in until the 90^0 bend is pressed up against the surface of the foam. Place the protruding music wire into the hole in the jig and pull it though the jig until the foam is setting flush against the surface of the jig. The jig should now be holding the pushrod perpendicular to the foam displacer.

Press down on the wire at the center of the displacer to make a small depression in the foam no more than 1/32" (0.79 mm) deep. Fill the depression with high temperature epoxy to attach the pushrod to the displacer. Leave the displacer and pushrod in the jig until the epoxy has cured.

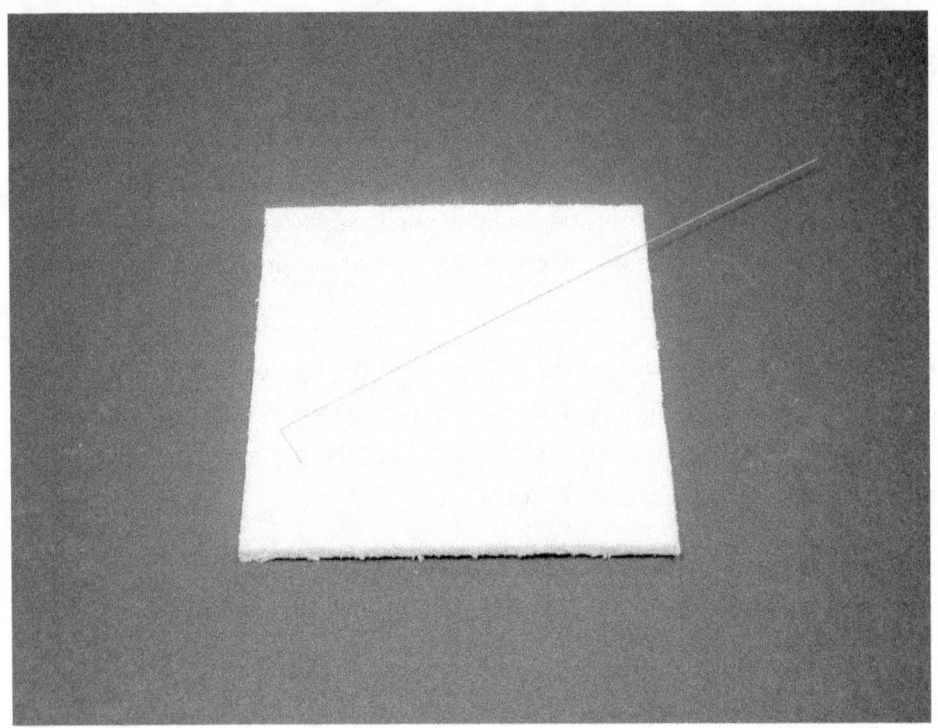

The displacer panel and the displacer pushrod are ready for assembly.

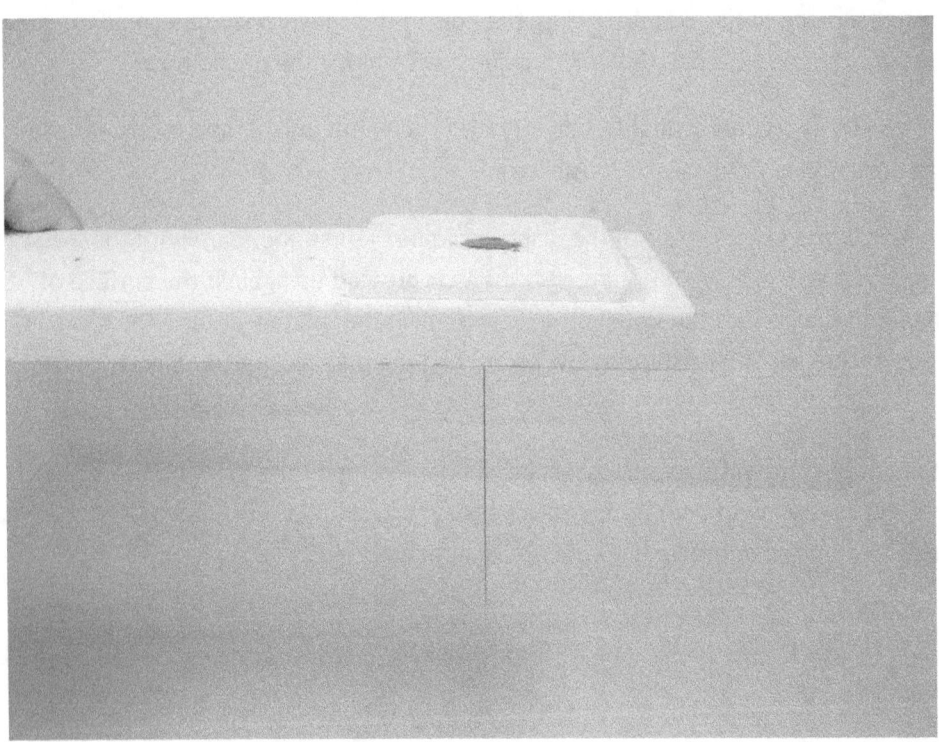

The displacer pushrod is held at the correct angle by the wooden jig. A hole is drilled through the board to hold the shaft at the correct angle. High temperature epoxy is used to attach the pushrod to the displacer panel.

Prepare the Drive Cylinder Tube

As noted earlier, the tubing used to make the drive cylinder is approximately 1" (25.4 mm) in diameter. Cut a piece of this tubing to a length of 5/8" (15.88 mm). Cut the pipe carefully to maintain a 90⁰ angle on each end of the short drive cylinder.

Use fine sand paper to remove the glossy factory surface from the PVC pipe. This will also remove any labels and markings and improve the look of the finished part. Sanding the outer surface of the drive cylinder will help hold the latex drive diaphragm in place.

Place the drive cylinder over the hole near the corner of the pressure chamber top plate and glue it in place with epoxy. Apply the glue on the inside edge of the drive cylinder tube. Applying the glue on the inside edge will hide the glue joint from view when the engine is fully assembled.

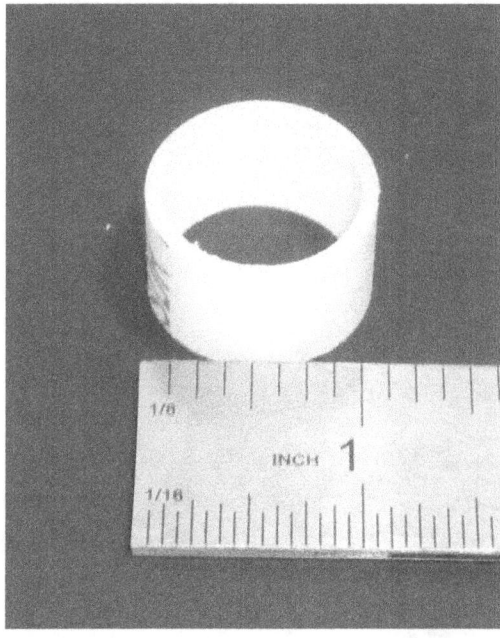

The drive cylinder is made from a short piece of 1"
(25.4 mm) diameter PVC pipe.

The drive cylinder is attached with epoxy. The glue is spread on the inside of the cylinder to create a nice looking airtight joint.

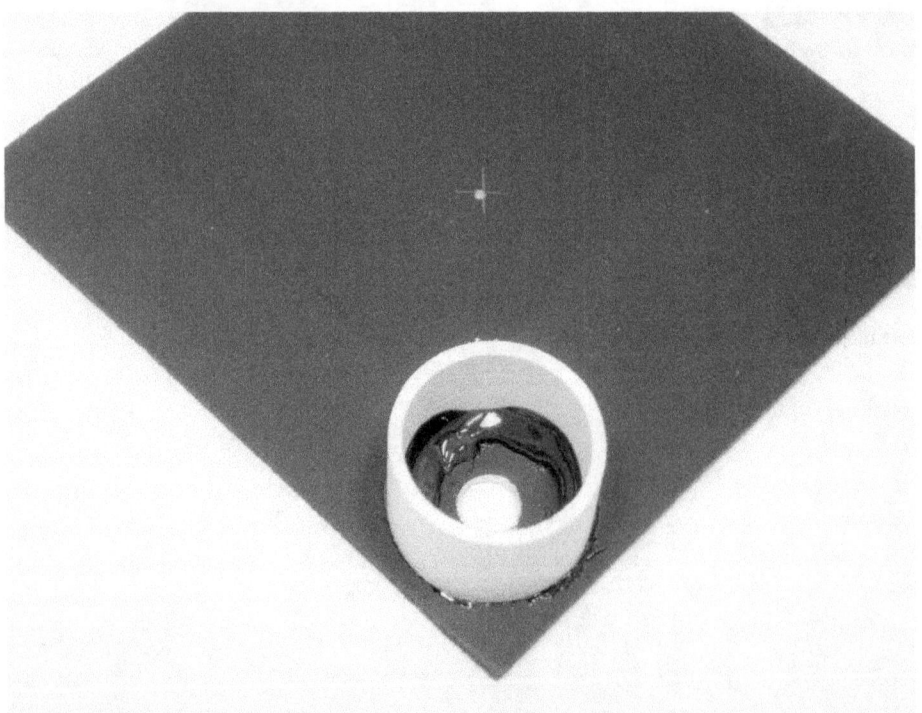

The drive cylinder is glued in place on the pressure chamber top. The glue is on the inside of the cylinder to provide an airtight seal that will be hidden from view after final assembly.

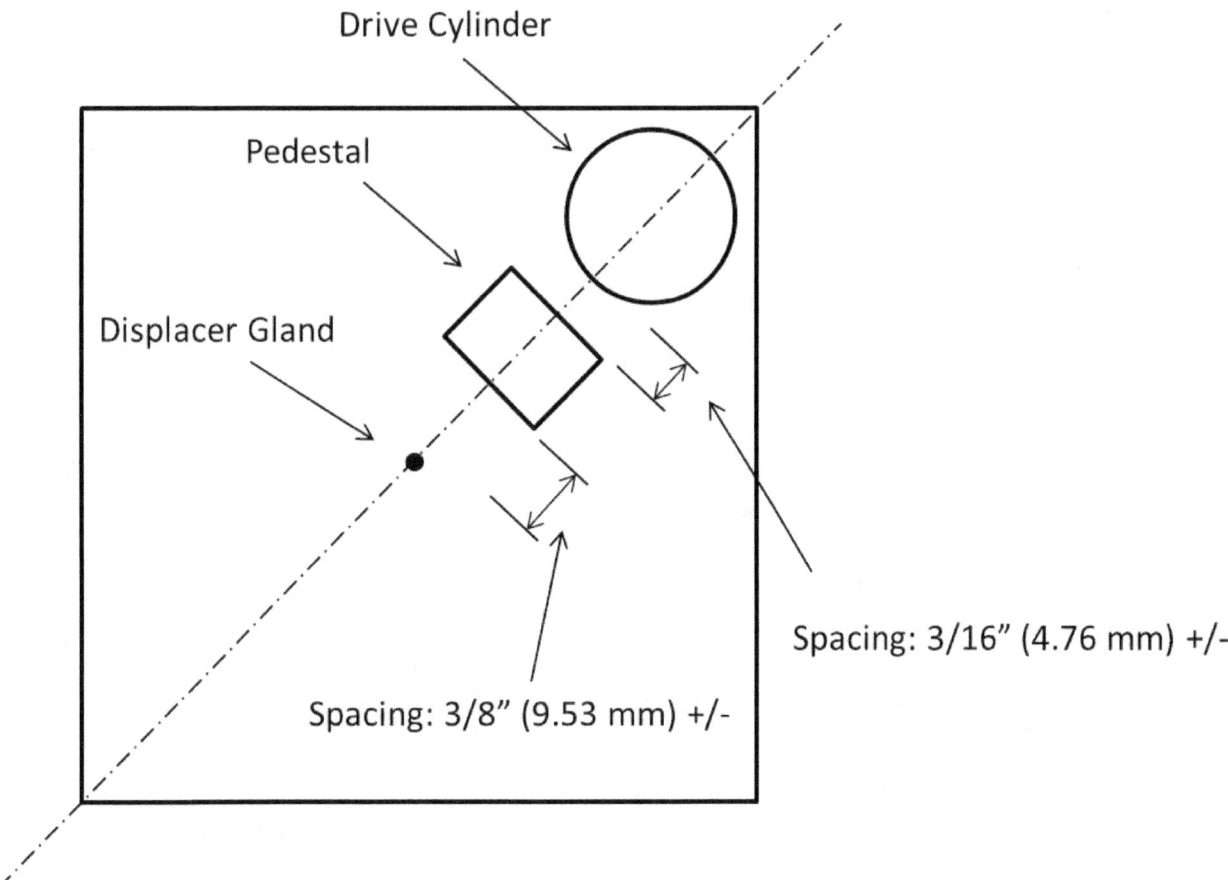

Drive Cylinder

Pedestal

Displacer Gland

Spacing: 3/16" (4.76 mm) +/-

Spacing: 3/8" (9.53 mm) +/-

This diagram illustrates the position of the parts on the top of the small square engine.

Attach the Displacer Gland

In the world of machinery, a "gland" is a sleeve within a stuffing box, fitted over a shaft in such a way as to prevent leakage of fluid while allowing a shaft or stem to move. Additionally, the gland of a Stirling engine allows the shaft to move with little or no friction. The gland is nearly air-tight. A tiny (very tiny!) pressure leak is desirable and helps the engine adjust for pressure changes as it warms up.

The gland is made of #24 (0.022" ID) Teflon tube. This provides a friction free fit for the 0.015" displacer pushrod.

Cut a piece of this Teflon tube to a length of 7/16" (11.11 mm).

Cutting Teflon tube requires a very sharp knife or razor blade, and a piece of 0.015" music wire. Insert the music wire into the Teflon tube before cutting. Place the tubing and wire on a hard flat surface. Press the blade of the knife against the tubing and roll the tubing on the flat surface until the knife has cut all the way around the tube. The wire in the middle of the tube will stop the knife from cutting all the way through the tube unless the tube is rolled. This will prevent the end of the tube from being crushed or deformed during the cutting process.

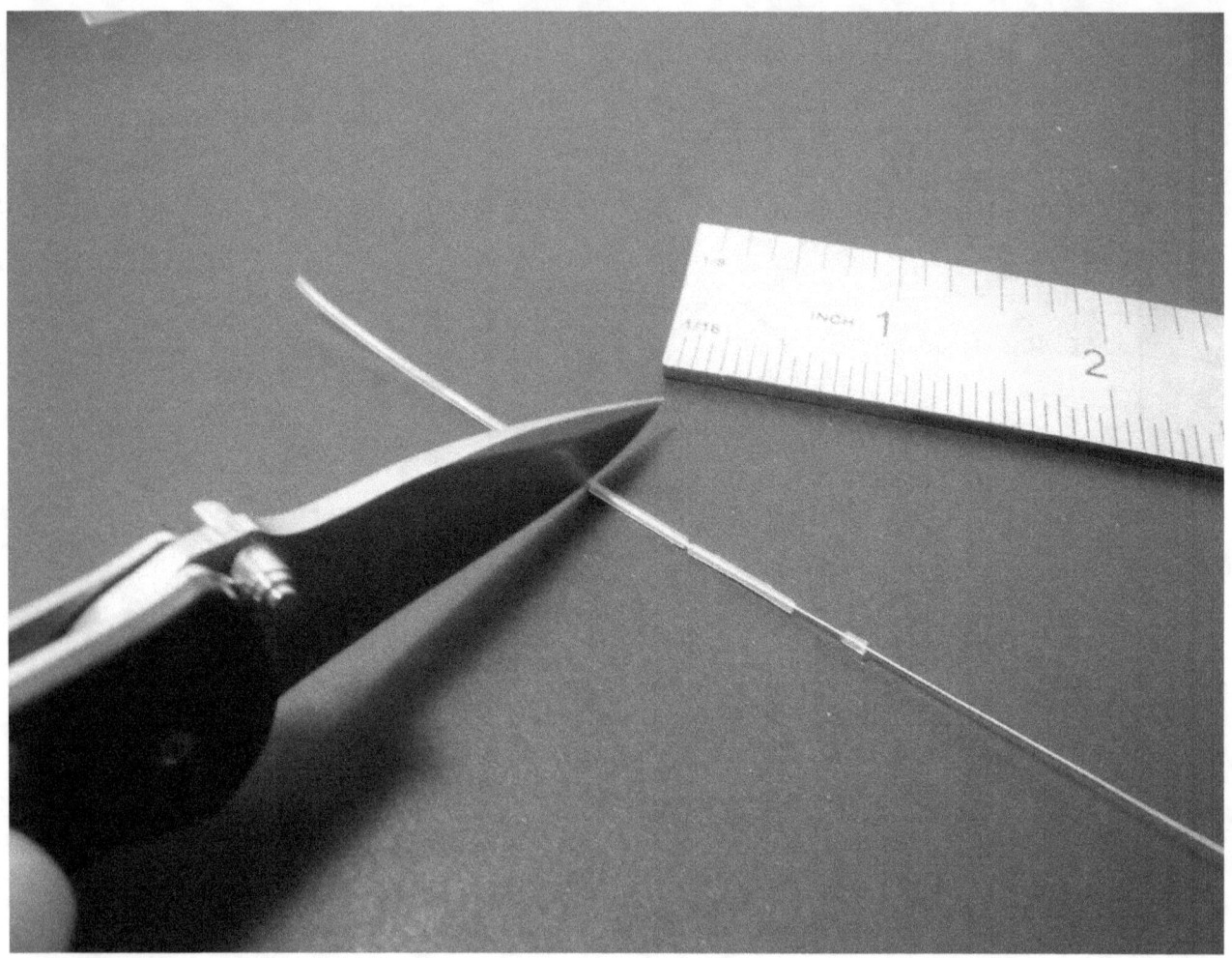

Insert a piece of music wire inside the Teflon tube before cutting. Roll the tube on a hard surface to cut around the wire. Cutting in this manner prevents the tube from deforming.

To attach the displacer gland:

1. Locate the displacer/pushrod assembled previously, the pressure chamber top plate, and the Teflon gland.
2. Slide the displacer pushrod through the top plate of the pressure chamber.
3. Set the displacer/top plate on a flat level surface with the pushrod pointing upward.
4. Slide the 7/16" (11.11 mm) gland tube onto the displacer pushrod.
5. Test for a friction free fit by verifying that the short piece of Teflon tube can fall under its own weight when dropped.
6. Let the Teflon gland drop inside the hole on the pressure chamber top plate until the end of the Teflon tube is flush with the inside surface of the top plate.
7. Apply a small bead of epoxy glue around the base of the gland tube, sealing it to the pressure chamber top plate. Take care that the glue does not touch the displacer pushrod. Allow the glue to cure.

The pressure chamber top plate is placed over the displacer with the pushrod in place. The Teflon tube gland is shown here before glue is applied.

High temperature epoxy is applied around the base of the displacer gland.

Create the Flywheel

The flywheel is the largest and most visible moving part on this engine.   If this part is built with care it will give the engine a very nice finished look. The goal is to make a flywheel that is perfectly round with a connection point for the axle that is squared to the flywheel and in the exact center. If the end result is less than perfect, the engine will still operate, and may even operate quite well. It just looks better if it is round, squared to the shaft, and centered.

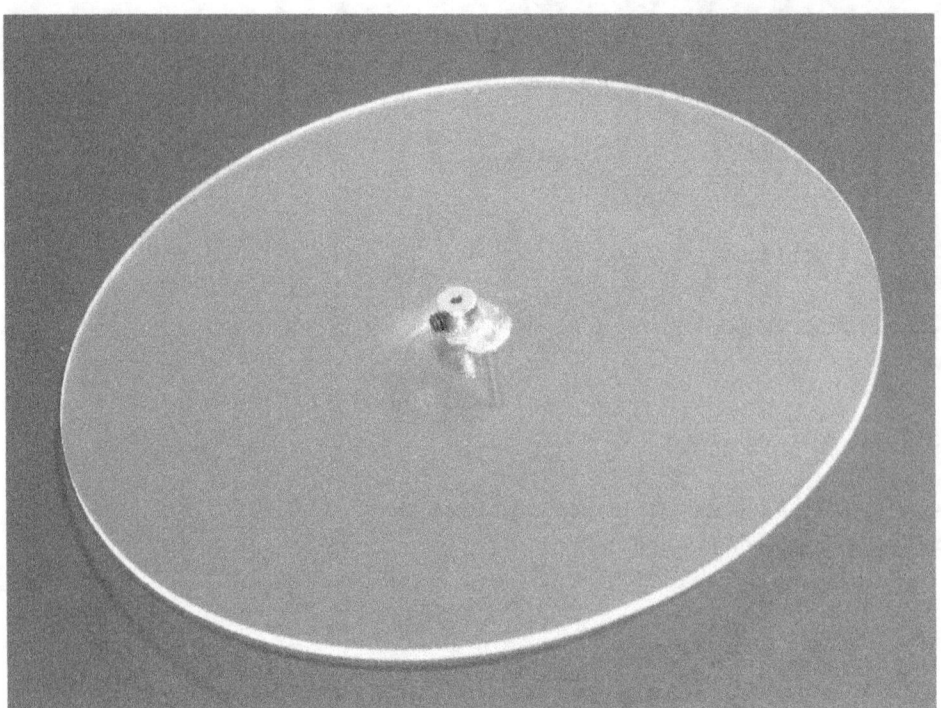

The flywheel pictured here was cut on a band saw using a simple homemade circle cutting jig.

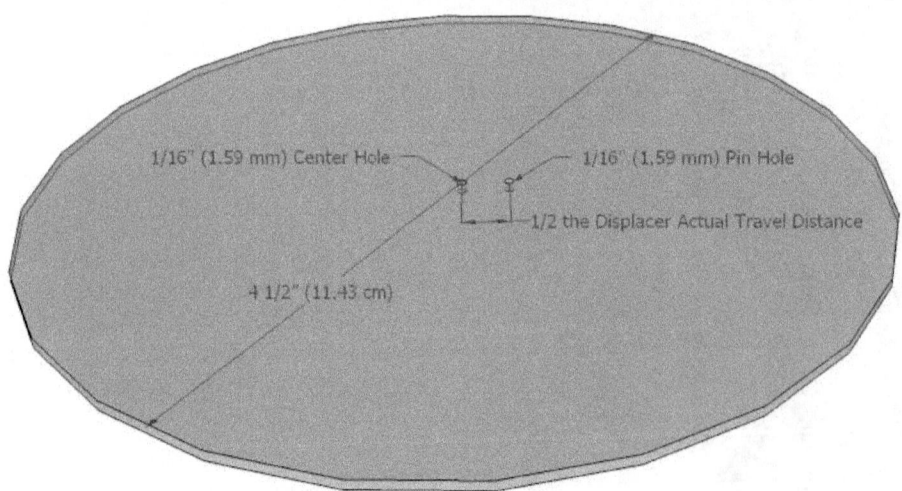

1/16" (1.59 mm) Center Hole

1/16" (1.59 mm) Pin Hole

1/2 the Displacer Actual Travel Distance

4 1/2" (11.43 cm)

The flywheel for the small engine has a diameter of 4 1/2" (11.43 cm).

48

Make a Simple Circle Cutting Jig

The jig can be made from a flat piece of scrap lumber. Dimensions are approximately 3 1/2" (8.89 cm) wide by 3/4" (19.05 mm) thick and about 16" (40.64 cm) long.

Drill a 1/16" (1.59 mm) hole about 5" (12.7 cm) from one end of the board, about 1 1/2" (38.1 mm) from the side of the board that will be facing the band saw blade. Insert a piece of 1/16" (1.59 mm) music wire in the hole, the same type of music wire that will be used to make the axle.

Draw a line from the pin to the edge of the board. Draw another line from the pin that is parallel to the edge of the board. This is the depth of cut mark. Draw a third line to mark the desired radius dimension of the flywheel. Make a saw cut from the front edge of the board to the depth of cut mark along the radius mark.

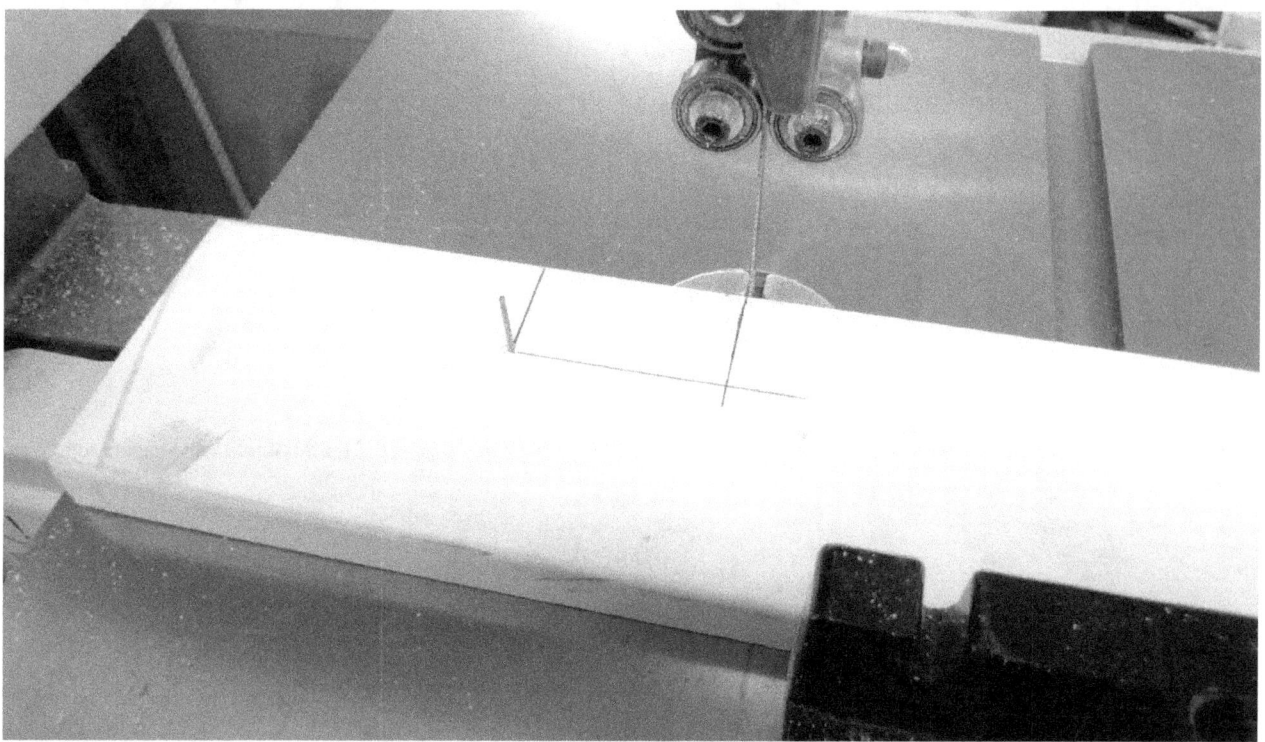

This is the circle cutting jig in place on the band saw. The center pin is on the left side. The saw blade is aligned with the radius mark. The horizontal line that runs parallel to the long edge of the board is the depth of cut mark.

Cutting the Flywheel with a Band Saw

Start with a piece of square acrylic sheet that is slightly larger than the diameter of the finished flywheel. Drill a 1/16" (1.59 mm) hole in the center of the acrylic sheet. Place the acrylic sheet on the circle cutting jig so that the pin on the circle cutting jig is going through the hole in the center of the acrylic sheet.

Hold the jig against the miter gauge and line up the blade with the cutting mark on the circle cutting jig. Turn the band saw on and press forward until the blade is even with the depth of cut mark, which is the same distance from the edge of the jig as the pin is.

Hold the jig still with one hand, and use the other hand to slowly rotate the acrylic around the pin, making a circular cut as you go. When the cut makes a complete circle, back the jig away from the blade and stop the band saw.

Drill a 1/16" (1.59 mm) hole in the center of the acrylic sheet. The hole is the same size as the material that will be used to make the axle.

Place the acrylic sheet on the circle cutting jig with the pin in the center hole.

Hold the acrylic sheet steady and move the jig forward to make the initial cut. Stop the forward movement of the jig when the blade reaches the depth of cut mark.

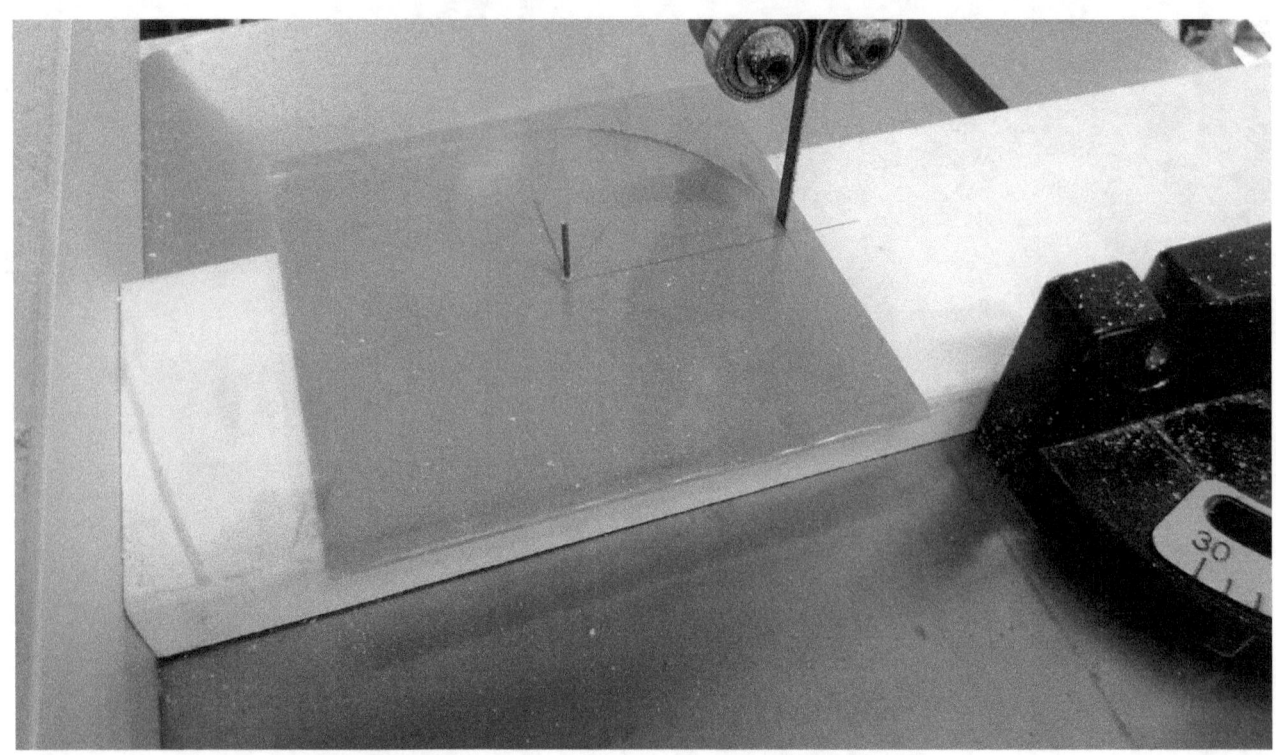

Hold the jig in place with one hand and slowly turn the acrylic sheet with the other hand.

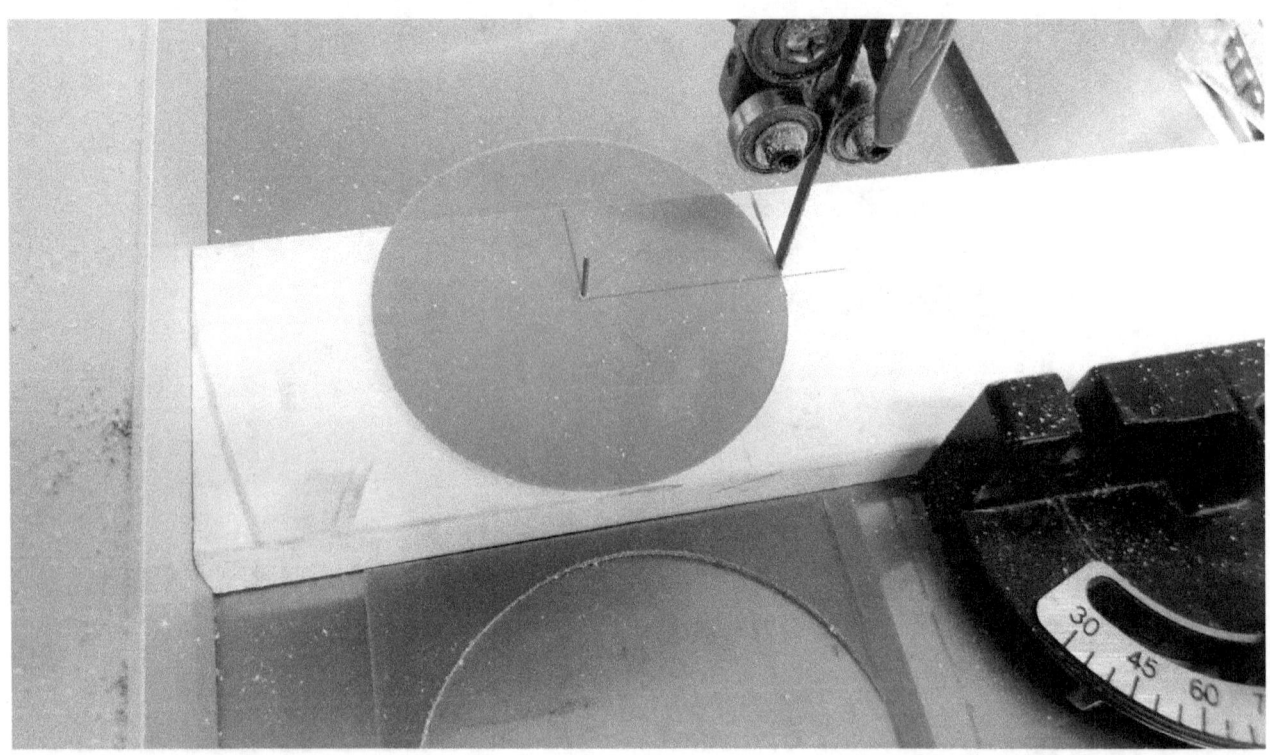

Continue turning the acrylic sheet until the circle cut is complete. The end result is a nice round flywheel with the axle hole in the center.

Cutting the Flywheel without a Band Saw

The flywheel can be cut with hand tools. Begin by marking the center point of the flywheel on the acrylic sheet. Use a compass point or another sharp object to make a small indentation at the center point of the wheel. With that point as the center, use a pair of dividers or a compass to mark the outer edge of the circle. Use a coping saw with a very fine blade (or a jeweler's saw) to carefully cut along the line.

Wrap a piece of 120 grit sandpaper around a small flat block of wood and use this sanding block to clean up the imperfections in the cut edge of the flywheel.

Finally, drill the center hole with a 1/16" (1.59 mm) drill bit.

Assemble the Pressure Chamber with the Displacer

Place the displacer pushrod through the gland and place the top plate on the pressure chamber with the displacer inside. Check the fit of all the parts and ensure that the displacer can move up and down without being obstructed. Make any necessary adjustments to the displacer before the top is glued onto the pressure chamber.

Use silicone adhesive to attach the pressure chamber top plate to the sidewalls. Spread a thin layer of the adhesive on the top edge of the sidewalls and smooth it with your finger. Use enough adhesive to create a seal, but no more. If too much adhesive is applied it may ooze into the inside of the pressure chamber and interfere with the motion of the displacer. Clamp the parts together with light pressure and set them aside until the silicone adhesive has cured.

Carefully glue the top plate onto the pressure chamber with silicone adhesive. Check carefully that the displacer can move up and down without obstruction inside the pressure chamber. Silicone adhesive is used for this joint because it can be opened up later if repairs are necessary.

Drill the Hole for the Crankshaft Pin in the Flywheel

The crankshaft pin is mounted in a hole that is drilled near the center hole of the flywheel. The spacing between the center flywheel hole and the crankshaft pin determines how far the displacer will travel as the flywheel rotates. The spacing between these two holes will be exactly half the _actual travel distance_ of the displacer.

Measure the travel of the displacer pushrod. To do this, take a measurement from the top of the pressure chamber to the top of the displacer pushrod when the displacer is on the bottom of the pressure chamber. Now lift up on the pushrod until the displacer is at the top of the pressure chamber and measure it again. Subtract the smaller number from the larger number to calculate the _total available travel distance_.

Subtract 1/16" (1.59 mm) from the _total available travel distance_ to get the _actual travel distance_. The actual travel distance will be slightly shorter than the total available distance so that the displacer will not touch the top or bottom of the pressure chamber as the engine runs. Shortening the travel distance by 1/16" (1.59 mm) will provide 1/32" (0.79 mm) of clearance above and below the displacer and prevent it from coming into contact with the pressure chamber.

It is important that the displacer does not come into contact with the pressure chamber as the engine is running. If the displacer hits the pressure chamber this will increase friction or prevent the engine from rotating freely. Also, it is good to keep a small cushion of air between the displacer and the pressure chamber top and bottom plates. The cushion of air reduces drag from what some refer to as "pull-off friction."

Pull-off friction can be demonstrated by holding a flat piece of cardboard against the ceiling with a broom handle. If the broom handle is quickly removed the cardboard does not immediately drop. The air pressure on the bottom of the cardboard holds it up until air is able to get in between the cardboard and the ceiling and equalize both pressures. This same effect can happen inside the pressure chamber if the displacer comes to rest in contact with the top or bottom plate of the pressure chamber. For this reason the engine is designed so that the displacer clearance is at least 1/32" (0.79 mm).

Once you have determined the _actual travel distance_ of the displacer, divide that distance in half. This number will be the distance between the center hole of the flywheel and the hole for the crankshaft pin. Drill the hole for the crankshaft pin with a 1/16" (1.59 mm) drill.

The same circle cutting jig that was used to make the flywheel on the band saw is also an excellent jig for measuring and drilling the hole for the displacer crankshaft pin. The illustration shows how a pair of dividers can be used to measure the distance from the center pin to find the location for the displacer crankshaft pin hole.

Set the dividers to the distance needed for the offset of the crankshaft pin. Center one divider point on the center pin and the other point on the middle of the drill bit. Drill the hole when the alignment is correct.

Attach the Shaft Collar to the Flywheel

The flywheel is attached to the axle by means of a small round shaft collar that contains a set screw. The shaft collar is glued to one side of the flywheel in the exact center. Great care needs to be taken to align the shaft collar so that the axle will be perpendicular to the surface of the flywheel.

An alignment jig similar to what was used to attach the displacer pushrod will help attach the shaft collar with proper alignment. Use a drill press to drill a 1/16" (1.59 mm) hole though a flat board. This board will serve as the alignment jig. Place a short piece of 1/16" (1.59 mm) music wire through the hole in the board. The music wire needs to be long enough to pass all the way through the board, the flywheel, and the shaft collar.

Place the flywheel over the music wire and press it flat against the surface of the board. Place the shaft collar over the music wire and press it flat against the flywheel. If the dry fit looks good, lift the shaft collar and spread a small drop of 5 minute epoxy under the shaft collar and press it against the surface of the flywheel.

The music wire will have to be removed before the epoxy is completely cured or it may become permanently attached. When the epoxy begins to harden, pull the music wire down through the alignment jig from the back side. This should leave the shaft collar glued over the center hole of the flywheel in near perfect perpendicular alignment.

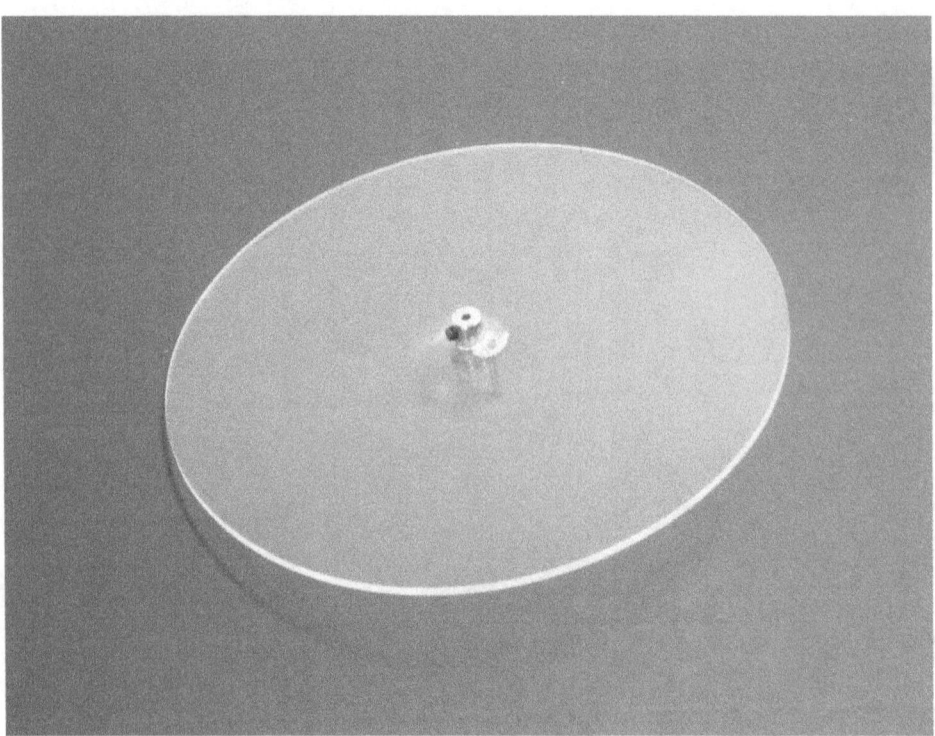

The shaft collar is attached over the center hole of the flywheel.

Make and Attach the Displacer Crankshaft Pin

There are two crankshaft pins. The displacer crankshaft pin attaches to a hole in the flywheel. The drive crankshaft pin is on the opposite end of the axle, over the drive diaphragm.

The displacer crankshaft pin is made from 0.0625" music wire, and is 1/2" (12.7 mm) long. The surface of the music wire should be smooth, with no gouges or tool marks, and it must be straight. Mark the wire where it is to be cut, clamp it in a vise, and then use the corner of a file to score the wire at the mark on two sides. Padding the vise and the pliers with paper will reduce the chances of scratching the music wire. Carefully bend the wire with pliers and it will break at the scored mark. Use a file or sandpaper to smooth the ends of the pin.

Glue the pin into the displacer pin hole on the flywheel. The location of the hole was calculated earlier. Use a small amount of epoxy to attach the pin to the flywheel. The pin must be perpendicular to the surface of the flywheel.

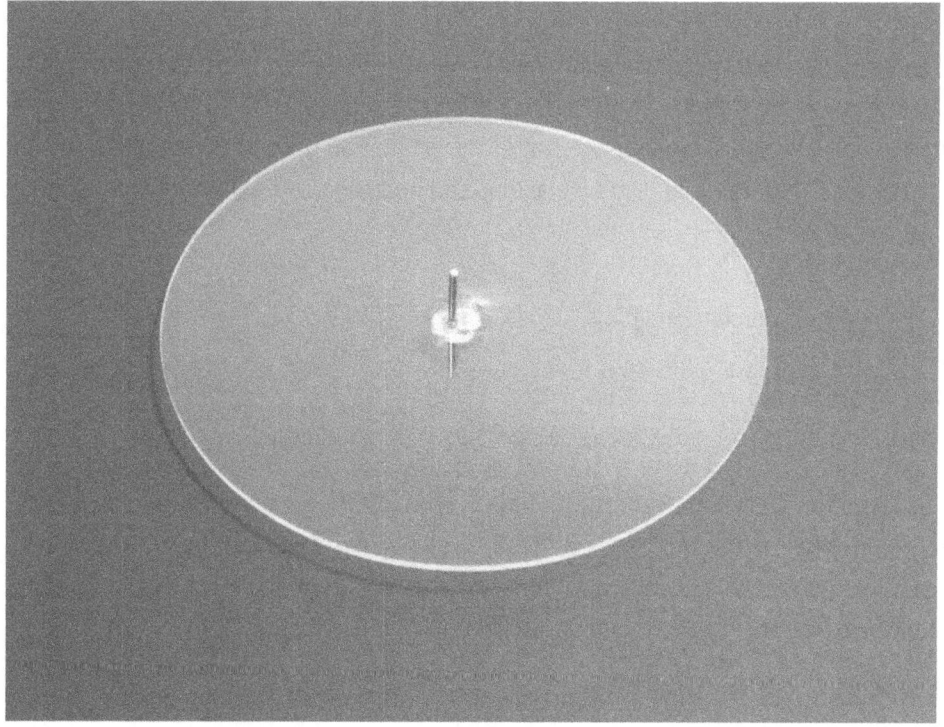

The crankshaft pin is attached to the flywheel on the opposite side as the shaft collar. The crankshaft pin is in the hole that is slightly off center.

Make the Main Axle

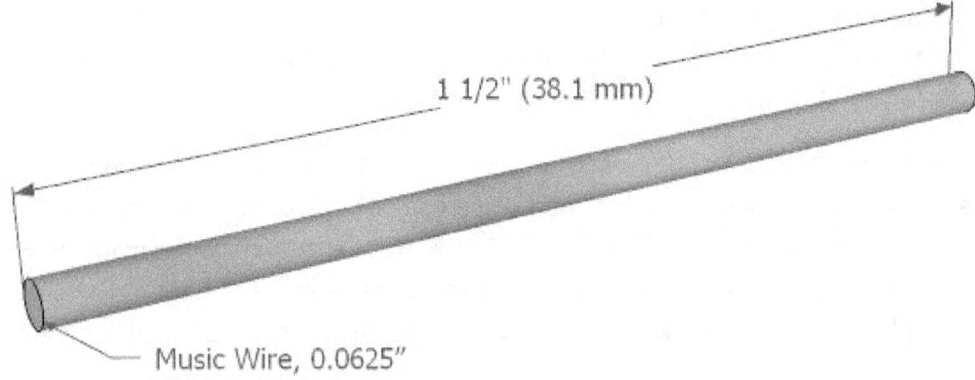

1 1/2" (38.1 mm)

Music Wire, 0.0625"

The axle is also made of 0.0625" music wire cut to a length of 1 1/2" (38.1 mm). Measure, score, and break the axle wire using the same technique that was used to cut the displacer crankshaft pin. Use a file or sandpaper to smooth the ends of the axle.

Make the Pedestal

The pedestal is a small post that attaches to the top of the pressure chamber. The top of the post is made wider by attaching a small piece of wood to one side. This is done to increase the stability of the main axle bushings, which are mounted in a small hole drilled near the top of the post. The base of the post is kept small so that it will fit in the space available between the displacer gland and the drive cylinder.

Cut two pieces of wood to these dimensions:

- 3/4" (19.05 mm) x 9/16" (14.29 mm) x 3" (7.62 cm)
- 3/4" (19.05 mm) x 3/4" (19.05 mm) x 1/4" (6.35 mm)

Attach the smaller piece to the wide side of the pedestal as illustrated. Use wood glue.

Drill a hole in the top of the pedestal for the axle. Position the hole 2 3/4" (69.85 mm) from the bottom of the pedestal (which is 1/4" (6.35 mm) below the top of the pedestal). The size of the hole must provide a snug fit for the Teflon tube used for the axle bushings. The tubing used here fits snugly in a 7/64" (2.78 mm) hole. Measure to verify the hole size needed for the bushing material that will be used.

The pedestal may be finished with paint or varnish to improve the appearance of the finished engine.

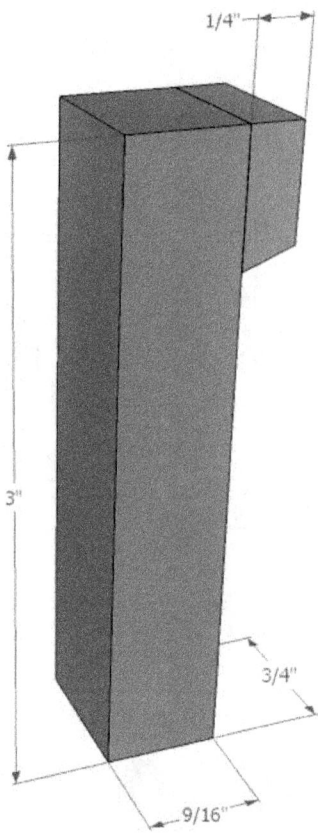

This image shows the assembled dimensions of the pedestal for Engine #4.

Cut and Mount Teflon Tube Axle Bushings

The axle bushings are made from AWG-14 (0.066") Heavy Wall Extruded Teflon Tube. Two pieces are required. Each piece is 3/16" (4.76 mm) long.

Cut the tubing by first placing a piece of 0.0625" music wire inside the tube at the place to be cut. Place the tubing on a flat surface and roll the tubing as you cut it with a sharp knife or razor blade. Cut down to the music wire as the tubing is rolled on the flat surface. Cutting in this manner prevents the tubing from being deformed during the cutting process. Cutting the tubing without the shaft inside can cause the tubing to flatten or kink at the point of the cut, causing friction in the bushing.

Insert the small pieces of Teflon tube into each end of the hole in the pedestal. Leave a small amount of tubing (about 1/16" (1.59 mm) or less) protruding from the hole on both sides. No gluing will be required if the hole is the correct size.

The outside diameter of the axle is 0.0625", which allows it to turn freely inside the 0.066" Teflon tube. Insert a piece of axle material into the bushings to test the fit. Make any adjustments necessary to enable the axle to spin freely in the bushings.

The pedestal is shown here with the Teflon bushings ready to be installed.

The Teflon bushings have been installed. Note that they protrude slightly above the surface of the pedestal.

Dry-Assemble and Measure for Locating the Pedestal

Attach the axle to the flywheel using the shaft collar, and insert this assembly into the bushings of the pedestal. The flywheel will be on the straight side of the pedestal.

Position the pedestal on top the pressure chamber between the drive cylinder and the displacer pushrod. The axle must be centered over the displacer pushrod and the drive cylinder. The pedestal must be positioned so that both the pushrods can be attached to their respective pins at a 90° angle to the axle. Once the ideal position for the pedestal has been found, mark the position with a pencil. Set the pedestal aside for now. It will be attached after the drive crankshaft has been assembled.

Make the Drive Crankshaft

The drive crankshaft plate creates the offset for the crankshaft that attaches to the drive diaphragm. It holds the drive crankshaft pin parallel to the axle with an offset of 3/32" (2.38 mm).

Draw the shape of the crankshaft plate on a piece of aluminum stock and drill the hole before cutting the plate. The plate is made from 0.062" aluminum sheet, the same material recommended for the top and bottom plates of the pressure chamber. The dimensions of the plate are 1/4" (6.35 mm) wide by 3/8" (9.53 mm) long. Drill a 1/16" (1.59 mm) hole at a position 1/8" (3.18 mm) from one end of the plate. Cut the small plate from the aluminum sheet after the hole has been drilled.

Cut the drive crankshaft pin from a piece of 0.062" music wire. The length of the pin is 1/2" (12.7 mm). Attach the pin to the hole in the plate with epoxy. The pin must be perpendicular to the surface of the plate. The pin is mounted on the front side of the plate.

Place a mark on the back side of the plate that is 3/32" (2.38 mm) away from the center of the pin. This mark will be used to position the shaft collar to the back side of the plate. Use epoxy to attach a 1/16" (1.59 mm) shaft collar to the back side of the plate. Take care that the set screw of the shaft collar does not become fouled with epoxy.

The crankshaft plate and the shaft collar are ready to be joined together with epoxy.

The crankshaft pin, plate, and shaft collar have been assembled.

Attach the Pedestal to the Pressure Chamber Top

The pedestal needs to be mounted so that it is aligned well with the displacer pushrod and the drive cylinder. Place a piece of straight music wire through the axle bushings to help to align the pedestal. This will make is easier to visually check the alignment and confirm the previous marks. Glue the pedestal in place with epoxy once the correct position is confirmed.

The pedestal is attached to the pressure chamber top. There is enough room for the flywheel between the pedestal and the displacer pushrod, and the drive crankshaft pin extends over the middle of the drive cylinder.

Test the Travel Distance of the Displacer

Install the axle, flywheel, and drive crankshaft on the pedestal. Position the flywheel so that the displacer crank pin is at the bottom position of its rotation. Use a marker or a piece of tape to make a reference mark on the displacer pushrod at the point where it comes in contact with the crank pin. Now rotate the flywheel until the pin is at the top of its rotation and raise the displacer pushrod until the mark is once again even with the pin. There should be about 1/16" (1.59 mm) free space between the displacer panel and the top of the pressure chamber when the displacer crank pin is at the top of its rotation. If it appears that the displacer will be able to move up and down without impacting the top or bottom of the pressure chamber, proceed to the next step. If the displacer crank is moving the displacer too far and it is impacting the engine, correct the problem by relocating the crank pin closer to the center of the flywheel.

Trim the Displacer Pushrod

Allow the displacer panel to rest on the bottom of the pressure chamber. Measure up from the top of the pressure chamber 1 1/2" (38.1 mm) and trim the displacer pushrod at this point.

Create the Displacer Connecting Rod and Teflon Bushing

The displacer connecting rod completes the connection between the top of the displacer pushrod and the displacer crank pin on the flywheel. A flexible connection is made to the displacer pushrod with two thin pieces of duct tape. The top end of the connecting rod is a piece of Teflon tube that will slip over the displacer crank pin on the flywheel. The displacer connecting rod is made from 0.015" music wire, which is the same size as the displacer pushrod.

Cut a piece of AWG-14 heavy wall extruded Teflon tube to a length of 7/16" (11.11 mm). Use the same cutting method described earlier so that the tubing does not become deformed at the cut.

Find or cut a piece of 0.015" music wire that is at least 3 1/2" (88.9 mm) long to construct the connecting rod. The exact length is not critical because it will be trimmed to fit. It may be easier to work with a longer piece and trim it to length after the bends are completed.

Place the short piece of Teflon tubing on a piece of axle stock (0.062" music wire) to help keep it straight while wrapping the connecting rod wire around the tubing. Make at least 1 1/2 wraps around the tubing with the connecting rod wire. Adjust the connecting rod wire so that it is at a 90° angle to the Teflon tube. The connecting rod wire and the Teflon tube should form the shape of a "T." Make the wraps tight enough so that they grip the Teflon tube and it does not fall out. Trim any excess wire when finished.

The Displacer Connecting Rod is fashioned from a length of music wire that is wrapped around a short piece of Teflon tube.

Attach the Displacer Connecting Rod

The connecting rod can now be installed between the flywheel and the displacer pushrod. Place the Teflon tube over the drive crankshaft pin on the flywheel. With the displacer panel resting on the bottom of the pressure chamber and the flywheel pin in its lowest position, trim the length of the connecting rod so that there is a gap between the end of the displacer pushrod and the end of the connecting rod. The gap should be between 1/16" (1.59 mm) and 1/32" (0.79 mm).

Cut two small pieces of duct tape to 1/8" (3.18 mm) x 1/2" (12.7 mm). Lift the displacer so that the end of the displacer pushrod is near the end of the connecting rod and fasten the two pieces together with the small pieces of duct tape. Position the duct tape so that the joint bends correctly in order to accommodate the motion of the flywheel.

Rotate the flywheel and observe the motion of the displacer panel. It should travel up and down inside the pressure chamber without touching the top or the bottom of the pressure chamber. Make any adjustments necessary so that the displacer does not come into contact with the top or bottom of the pressure chamber.

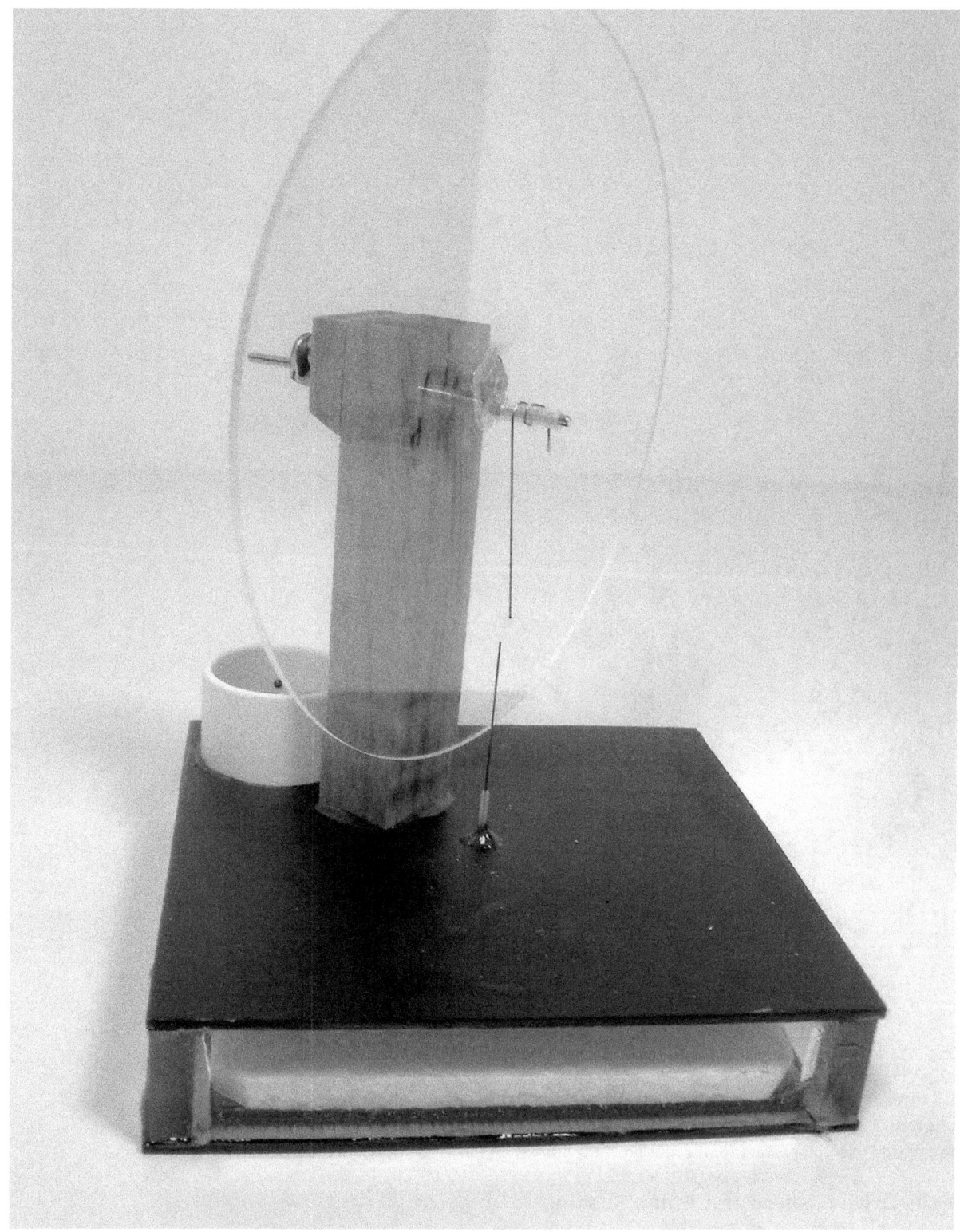

The displacer connecting rod has been trimmed to leave a small gap between the ends of the two rods.

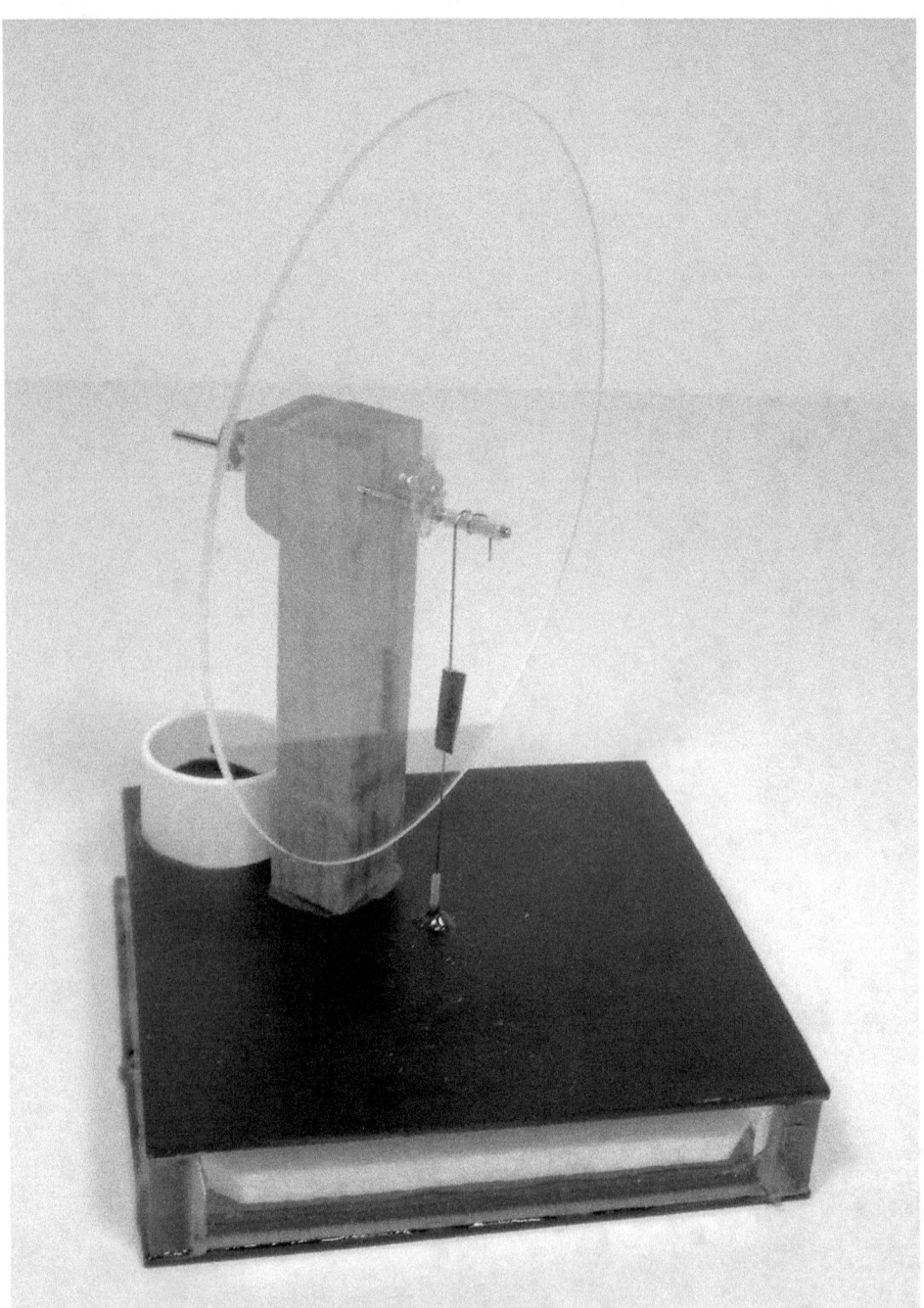

Two small strips of duct tape are used to connect the displacer pushrod to the displacer connecting rod. The flat surface of the duct tape is oriented so that is can act as a hinge that bends side to side to accommodate for the motion of the crankshaft pin as the flywheel rotates.

Create the Drive Pushrod and Teflon Bushing

The drive pushrod is made just like the displacer connecting rod was made, except there is no duct tape joint in the middle of the shaft. There is a Teflon tube at the top end of the pushrod. The Teflon tube rides on the crankshaft pin. The bottom end is bent into a loop that is folded over so that it mounts flat against the drive diaphragm.

Cut a piece of AWG-14 Teflon tube to a length of 7/16" (11.11 mm). Insert a piece of 0.062" music wire inside to help hold it while wrapping a piece of 0.015" music wire around it. Make at least 1 1/2 turns around the tubing, as before. Adjust the tubing so that it is held snugly at a 90⁰ angle to the pushrod.

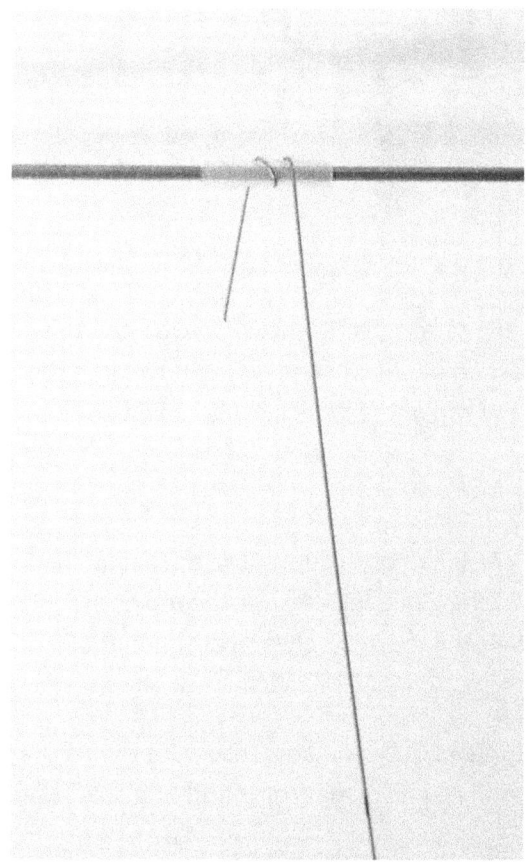

In this picture, the Teflon tube has been placed on a piece of axle wire and the pushrod wire is wrapped around the tubing to attach it to the pushrod.

Measure the distance between the center of the axle and the top of the drive cylinder. This will be the finished length of the pushrod.

Make a 90⁰ bend in the pushrod using the measurement just obtained. The distance from the axle to the top of the drive cylinder should be the same as the distance from the Teflon tube to the 90⁰ bend.

Use needle nose pliers to make a loop at the bottom of the pushrod. The loop should be about 7/16" (11.11 mm) in diameter. The loop is made in such a way so that if the connecting rod was placed on a flat surface it could stand upright with the loop flat against the table top.

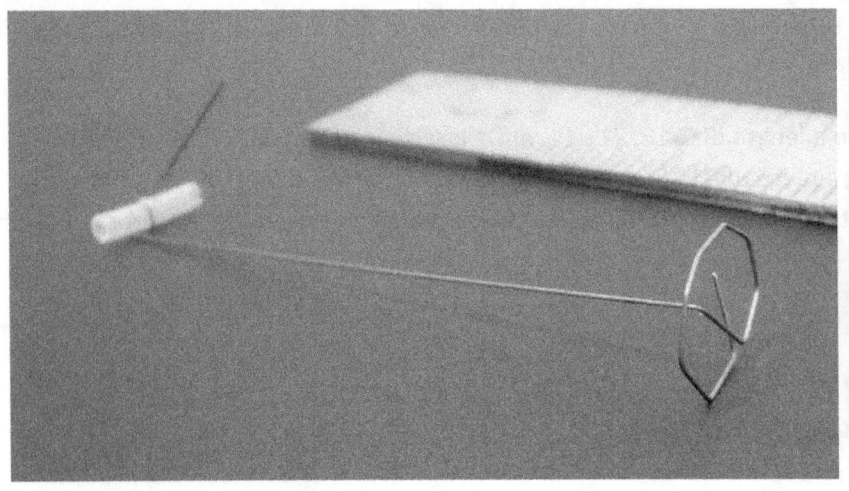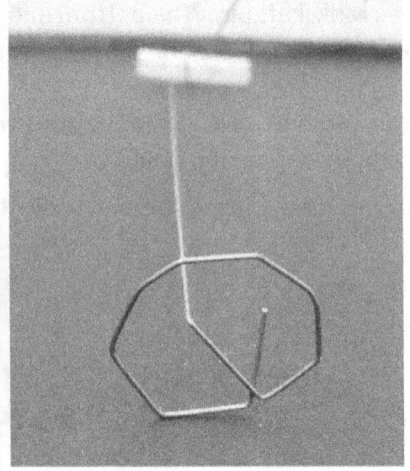

These pictures show the loop formed at the 90° bend in the pushrod. This loop is the attachment point for the drive diaphragm.

Verify the fit of the pushrod. Place it on the drive crankshaft pin and rotate the flywheel. The loop on the bottom should rise above the top of the drive cylinder at its highest point, and it should drop down into the drive cylinder at its lowest point. It should be close to even with the top of the drive cylinder when at its midpoint between high and low.

Create the Drive Diaphragm from a Latex Glove
The drive diaphragm is made from the fingertip of a latex glove. Wash and dry a latex glove so that all the powder is removed. Cut the fingers from the glove. Stretch one of the glove fingers over the top of the drive cylinder. Pull the latex down the outside of the drive cylinder until there is only a small amount of slack near the center of the drive diaphragm.

The drive diaphragm should stay in place without any help. If it appears to be slipping, secure it by placing a rubber band around the outside of the drive cylinder. A rubber band can be made from another glove finger if necessary.

Attach the Drive Pushrod to the Drive Diaphragm
The loop at the end of the drive pushrod should rest flat against the drive diaphragm and there should be no sharp wires threatening to puncture the diaphragm. Hold the loop against the center of the diaphragm and attach it using Superglue®.

This picture shows the drive pushrod attachment to the drive diaphragm. The wire pushrod is attached to the latex diaphragm with Superglue®. The diaphragm is made from the finger tip that has been cut from a latex glove.

Set the Crankshaft Timing Angle

Adjust the flywheel and the drive crankshaft so that there is a 90^0 offset between the displacer crank pin and the drive crank pin. The direction of the offset will determine the direction the motor rotates when running. The drive mechanism will follow 90^0 behind the motion of the displacer mechanism. That means when the displacer is all the way up at the top of the rotation, the drive mechanism will be halfway up. When the displacer is all the way down, the drive mechanism will be halfway down. Use the set screws on the shaft collars to hold the flywheel and the drive crankshaft in place.

Adjust the Drive Diaphragm Tension

The drive diaphragm should be adjusted so that there is just enough slack in the latex to allow the engine to rotate without stretching the material. If the material is too tight the engine will not run well because extra energy will be required to stretch the diaphragm. If the diaphragm is too loose the engine will not run well because the loose diaphragm will inflate and deflate without causing the crankshaft to move. Adjusting the tension of the drive diaphragm is one of the adjustments that can be made to fine tune the performance of the motor.

Check all the Connections

The engine should now be fully assembled and ready for its first run. Check all the connections by rotating the flywheel slowly and observing all the moving parts. Nothing should be falling apart when the flywheel is rotated. If the connecting rod or the pushrod becomes disconnected during this test, make adjustments so that they do not fall off.

Observe the motion of the displacer panel inside the pressure chamber and ensure that it does not impact the top or the bottom of the pressure chamber. It should move without any obstruction.

Run the Engine!

This Stirling Engine should run well over hot water. Fill a coffee cup or similar container with near-boiling water. Place the pressure chamber on top the cup of hot water. Allow it to warm up for 10 to 20 seconds. Turn the flywheel to start the engine.

The motor will continue to run as long as there is a temperature differential of 20° F (11° C) (or more) between the top and bottom surfaces of the pressure chamber. It may be possible to fine-tune the engine to operate on an even lower temperature differential with a little care and patience.

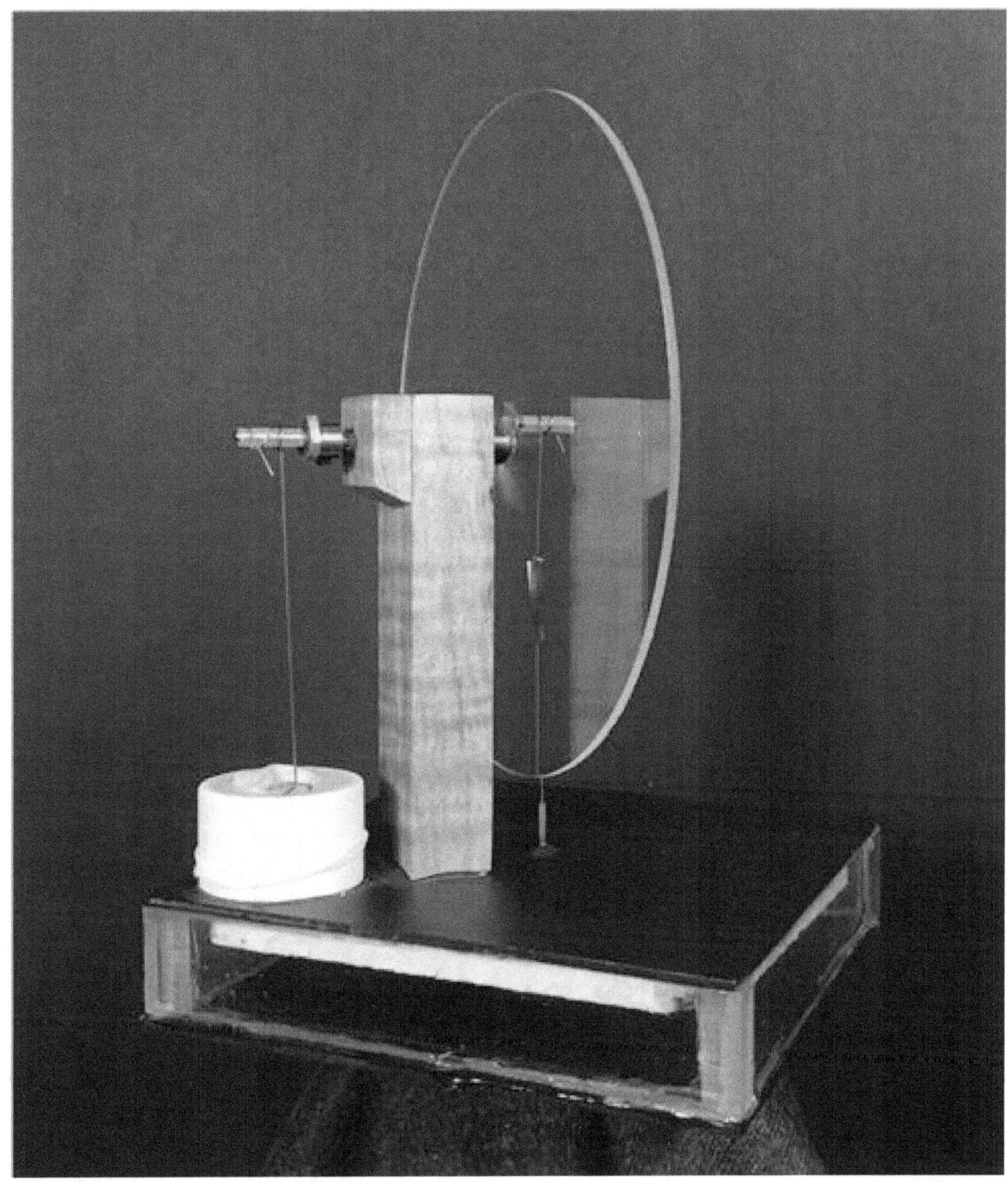

The finished 4" (10.2 cm) square engine.

Trouble Shooting Tips

As mentioned previously, the engine should run with a temperature differential of 20^0 F (11^0 C). Setting the engine on a cup of near-boiling water in a 70^0 F (21^0 C) room should provide a temperature differential of about 100^0 F (56^0 C). If the engine is not running under these conditions, there are one or more other problems that will need to be fixed, such as a small pressure leak or friction.

If the engine is not running well, it may be because of a problem in one of these four areas:

Temperature Differential: It may be possible to overcome a small pressure leak or friction by increasing the temperature differential. This will increase the power output of the engine and possibly overcome a small amount of friction or a small pressure leak.

Increase the temperature differential by adding ice to the top of the motor while the bottom is being warmed by the heat source. Do not attempt to add more heat, as this can damage the engine. The Styrofoam displacer material may melt if the heat source is too hot.

Pressure Leaks: It does not take much of a leak to prevent the engine from running well. There are a couple of ways to test for a pressure leak. The first method is to observe the behavior of the diaphragm when the engine is at running temperature.

Disconnect the drive diaphragm pushrod from the crank pin. Place the engine on a cup of hot water and wait a few moments for the bottom side to heat up. Now, rotate the flywheel so that the displacer rises and falls inside the pressure chamber and observe the motion of the drive diaphragm. The diaphragm should move up and down in response to the heating and cooling of the air inside the pressure chamber. If this motion is not present, or if it is very limited, there may be a pressure leak. It may also be possible that the tension of the diaphragm is too tight or too loose.

The other method for leak testing also involves removing the diaphragm pushrod from the crank pin. Once it is disconnected, pull upward on the pushrod for 5 to 10 seconds to inflate the diaphragm. Release the pushrod and observe the diaphragm. If it immediately deflates and returns to a low or neutral position, there may be a pressure leak.

Pressure leaks can happen at a number of places:

- Holes in the drive diaphragm
- Leaking around the edge of the drive diaphragm
- Leaking glue joints in the pressure chamber
- Leaking through excessive clearance around the displacer pushrod.

It may take a bit of detective work to find the leak. It may be possible to patch a small leak with a small drop of glue or silicone sealant, or it may be necessary to replace the defective part.

Friction: Small amounts of friction can have a huge impact on an LTD Stirling engine's ability to run. Friction occurs at every point where two moving parts touch, and at every point where a moving part contacts the atmosphere. In the micro-horsepower world of LTD Stirling engines, a tiny bit of friction can stop the engine from performing.

Check the rotation of the axle by removing the connections to the displacer and the drive diaphragm and spinning the flywheel. The flywheel should coast to a stop after about 30 seconds after receiving a good spin by hand. If the flywheel does not spin freely there is a problem with friction somewhere in the axle assembly. Locate the cause of the friction and repair the problem.

The flywheel rotation should be smooth and silent during the spin test. Vibration and noise are both indications of friction.

Crankshaft Timing: There must be a 90^0 phase difference between the two crank pins. This means that when one pin is in the 12:00 o'clock position, the other one is at either 9:00 o'clock or at 3:00 o'clock. The engine will run with a phase difference in either direction. The only difference will be the direction the engine rotates while running.

When the bottom of the engine is heated, the engine will run with the motion of the displacer moving ahead of the drive diaphragm. This means that when the displacer is at the top of the pressure chamber, the drive diaphragm is halfway up and moving in an upwards direction. As the flywheel rotates and the displacer comes to the lowest point in its travel, the drive diaphragm is halfway down and moving downward. It is important to know which way the engine will run so that the initial push (to get it started) is in the same direction.

Engine #5: Large Square Engine 6" (15.2 cm)

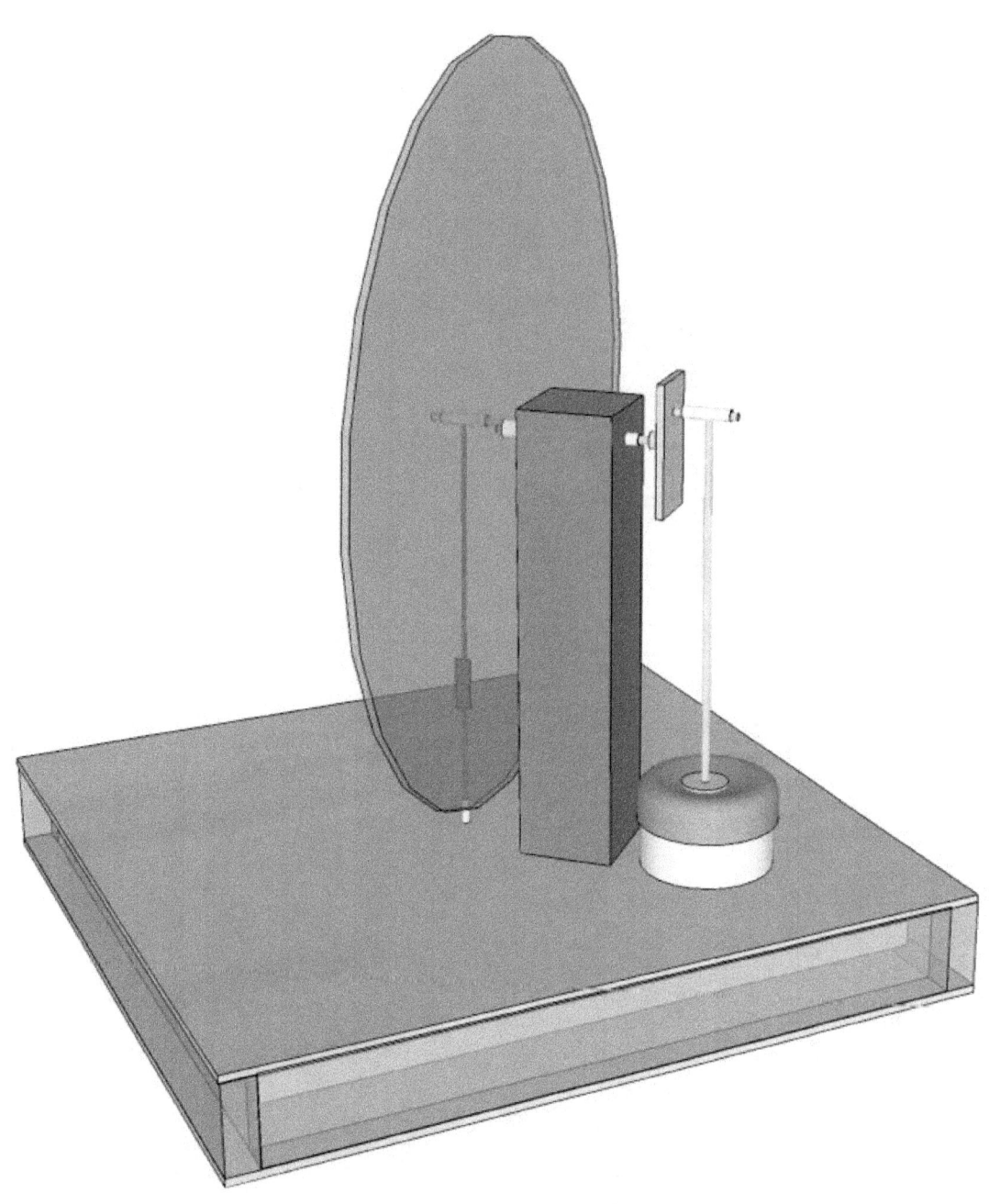

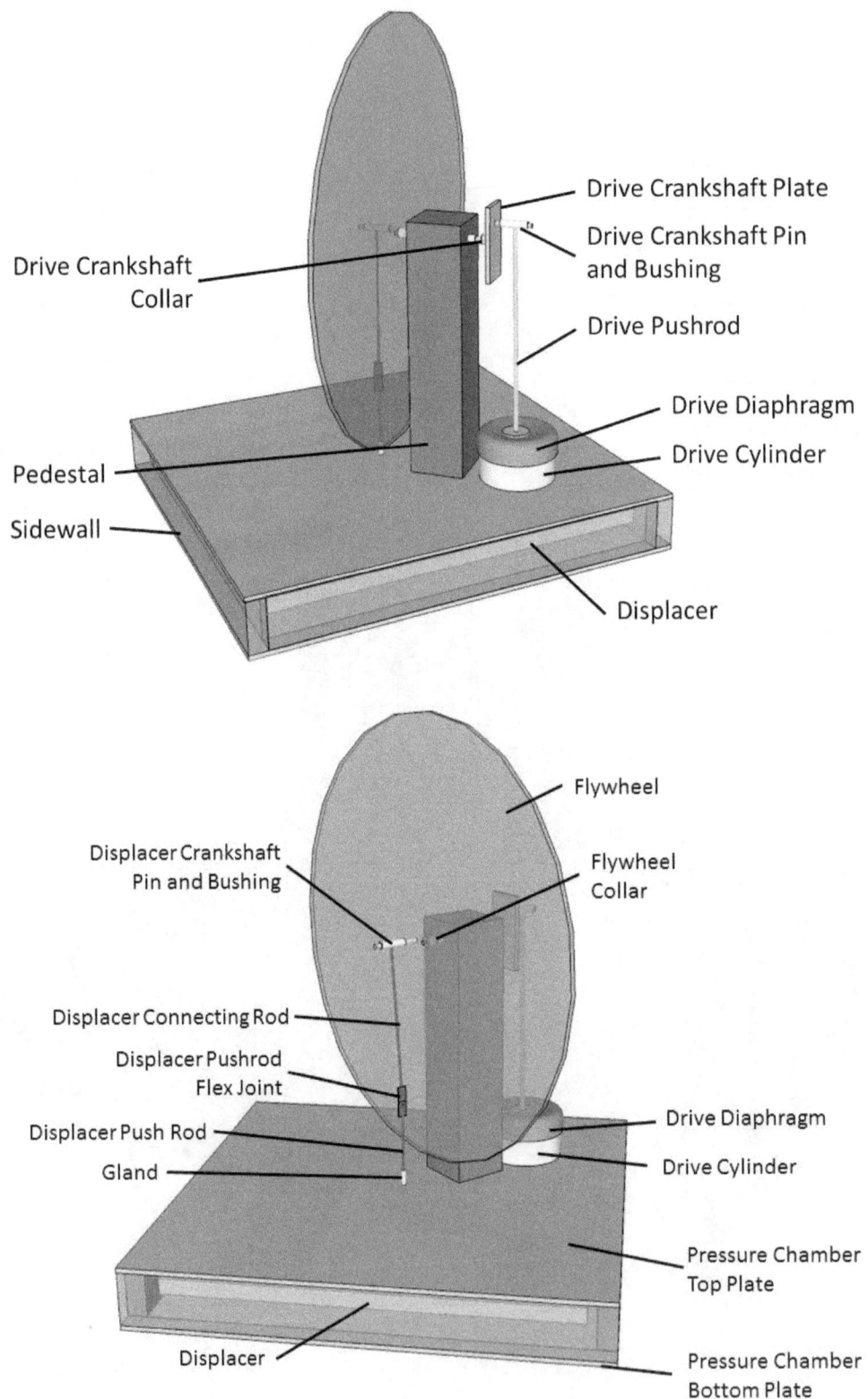

Drive Crankshaft Plate

Drive Crankshaft Pin and Bushing

Drive Crankshaft Collar

Drive Pushrod

Drive Diaphragm

Drive Cylinder

Pedestal

Sidewall

Displacer

Flywheel

Displacer Crankshaft Pin and Bushing

Flywheel Collar

Displacer Connecting Rod

Displacer Pushrod Flex Joint

Displacer Push Rod

Drive Diaphragm

Gland

Drive Cylinder

Pressure Chamber Top Plate

Displacer

Pressure Chamber Bottom Plate

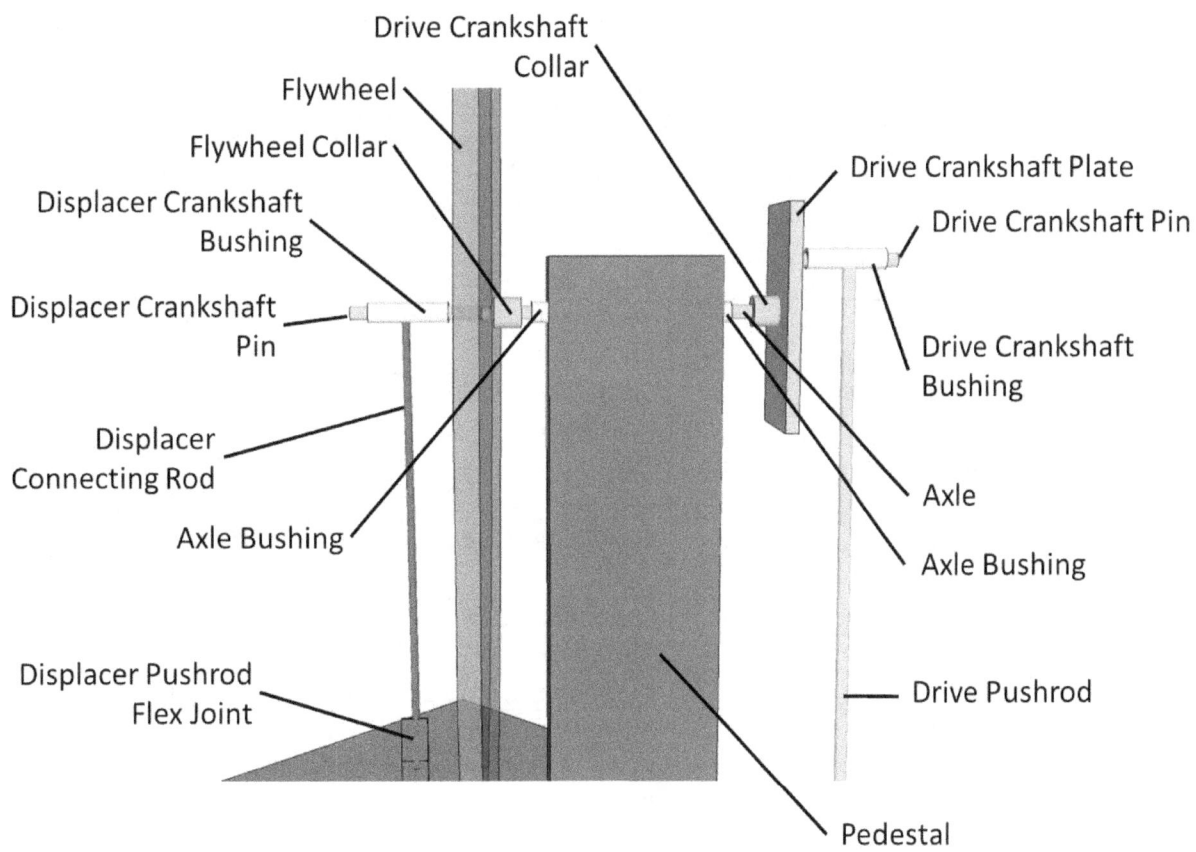

This image shows a close up view of the axle and drive assembly.

Parts and Materials

Part #	Part Name	Description
01	Flywheel	Clear Acrylic Sheet, round, 0.06" x 5 3/4" (14.61 cm) diameter*
02	Flywheel Collar	Shaft Collar, 1/16" (1.59 mm)
03	Pressure Chamber Bottom Plate	Aluminum Sheet, 0.062" x 6" (15.24 cm) x 6" (15.24 cm)
04	Pressure Chamber Top Plate	Aluminum Sheet, 0.062" x 6" (15.24 cm) x 6" (15.24 cm)
05	Displacer	Styrofoam, 0.20" (5 mm) x 5 1/4" (13.34 cm) x 5 1/4" (13.34 cm)
06	Displacer Crankshaft Pin	Music Wire, 0.0625 x 1/2" (12.7 mm)
07	Displacer Crankshaft Bushing	AWG-14 (0.066") Heavy Wall Extruded Teflon Tube, 3/8" (9.53 mm)
08	Long Sidewall (2 pieces)	Clear Acrylic, 1/4" (6.35 mm)** x 11/16" (17.46 mm) x 6" (15.24 cm)
09	Short Sidewall (2 pieces)	Clear Acrylic, 1/4" (6.35 mm)** x 11/16" (17.46 mm) x 5 1/2" (13.97 cm)
10	Pedestal	Wood, 3/4" (19.05 mm) x 3/4" (19.05 mm) x 3 1/2" (8.89 cm)
11	Displacer Connecting Rod	Music Wire, 0.015" x 4" (10.16 cm)
12	Displacer Pushrod	Music Wire, 0.015" x 6" (15.24 cm)
13	Displacer Pushrod Flex Joint	Duct Tape, 1/8" (3.18 mm) x 1/2" (12.7 mm), 2 pieces
14	Drive Crankshaft Pin	Music Wire, 0.0625 x 1/2" (12.7 mm)
15	Drive Crankshaft Bushing	AWG-14 (0.066") Heavy Wall Extruded Teflon Tube, 3/8" (9.53 mm)
16	Drive Crankshaft Collar	Shaft Collar, 1/16" (1.59 mm)
17	Drive Crankshaft Plate	Aluminum Sheet, 0.062" 1/2" (12.7 mm) x 1" (25.4 mm)
18	Gland	Teflon Tube, #24 (0.022" ID) x 7/16" (11.11 mm)
19	Drive Cylinder	PVC Pipe, 1" (25.4 mm) ID x 5/8" (15.88 mm)
20	Drive Diaphragm	Latex glove fingertip
21	Drive Pushrod	Music Wire, 0.015" x 4" (10.16 cm)
22	Axle	Music Wire, 0.0625"x 1 1/4" (31.75 mm)
23	Axle Bushings (2 pieces)	AWG-14 (0.066") Heavy Wall Extruded Teflon Tube, 3/16" (4.76 mm)

*Flywheel thickness can vary slightly from the prescribed thickness of 0.06", depending upon the material available to the builder. Acrylic sheet commonly sold in the US market for window glazing is slightly thicker (0.093") and is used for the flywheels of some of the engines in this book. Thin polystyrene sheet material used in covering framed pictures is only 0.05" thick and will suffice, but can be difficult to work with.

** It is also possible to make the sidewalls from slightly thinner 3/16" (4.76 mm) acrylic sheet.

Drawings and Dimensions

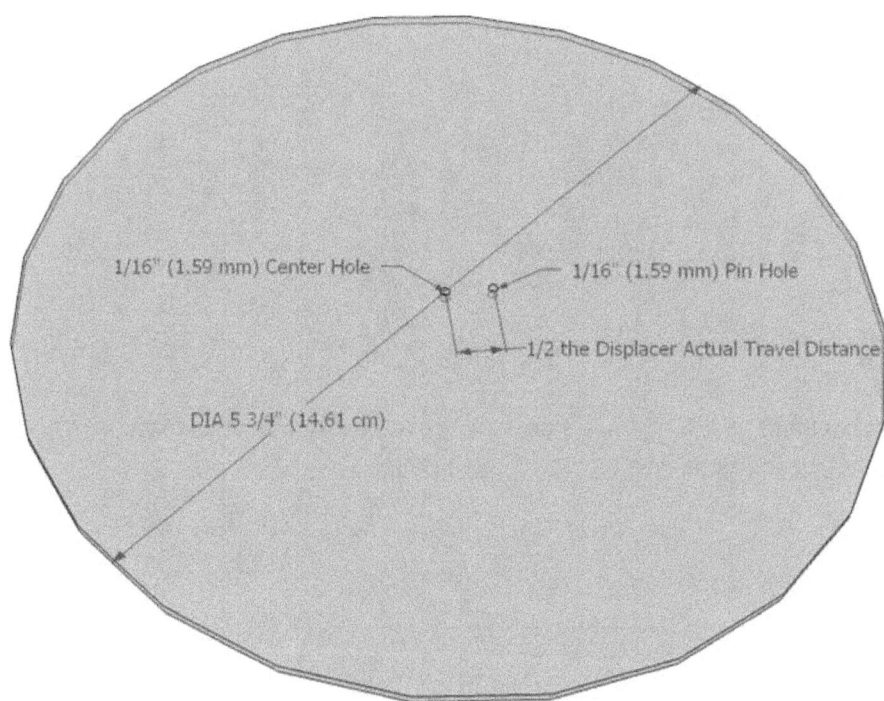

1/16" (1.59 mm) Center Hole

1/16" (1.59 mm) Pin Hole

1/2 the Displacer Actual Travel Distance

DIA 5 3/4" (14.61 cm)

Part # 01 Flywheel - Clear Acrylic Sheet, round, 0.06" x 5 3/4" (14.61 cm) diameter

Part # 02 Flywheel Collar - 1/16" (1.59 mm) Shaft Collar

Part # 16 Drive Crankshaft Collar - 1/16" (1.59 mm) Shaft Collar

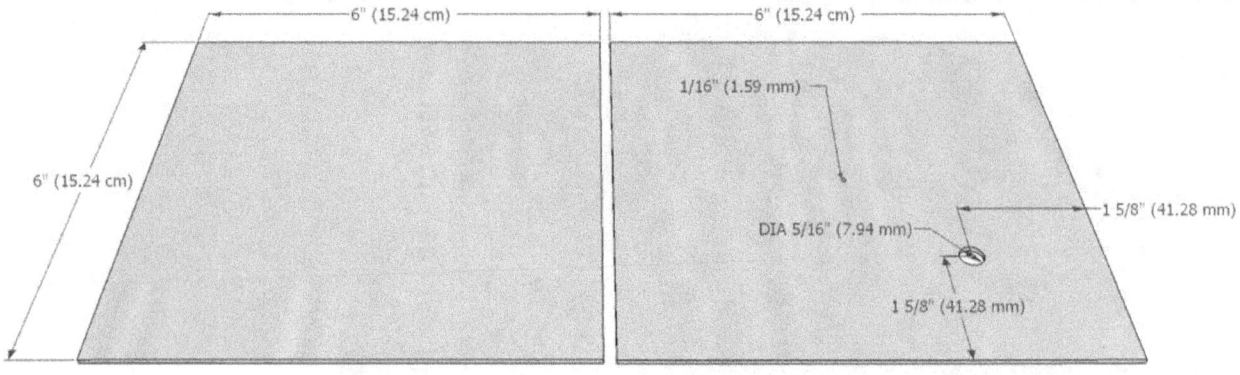

Part # 03 Pressure Chamber Bottom Plate　　**Part # 04 Pressure Chamber Top Plate**
Aluminum Sheet, 0.062" x 6" (15.24 cm) x 6" (15.24 cm) (2 pieces)

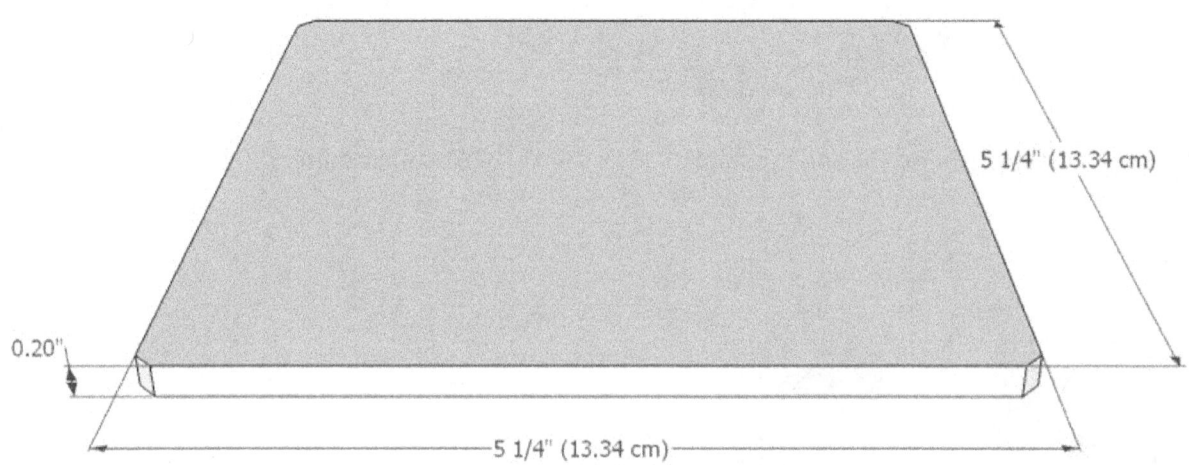

Part # 05 Displacer - Styrofoam, 0.20" x 5 1/4" (13.34 cm) x 5 1/4" (13.34 cm)

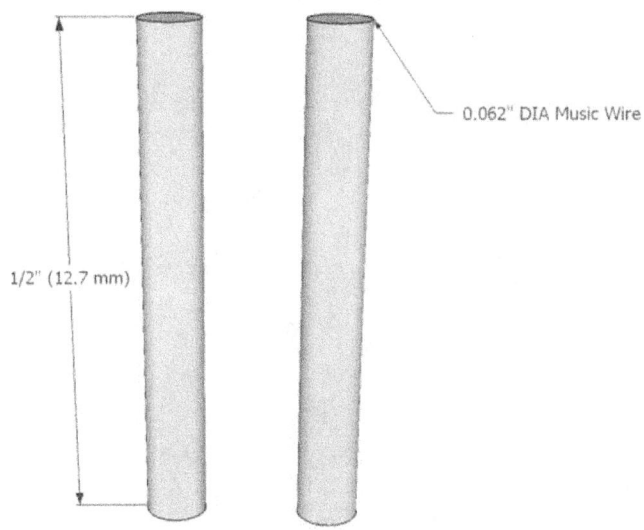

Part # 06 Displacer Crankshaft Pin - Music Wire, 0.0625 x 1/2" (12.7 mm)
Part # 14 Drive Crankshaft Pin - Music Wire, 0.0625 x 1/2" (12.7 mm)

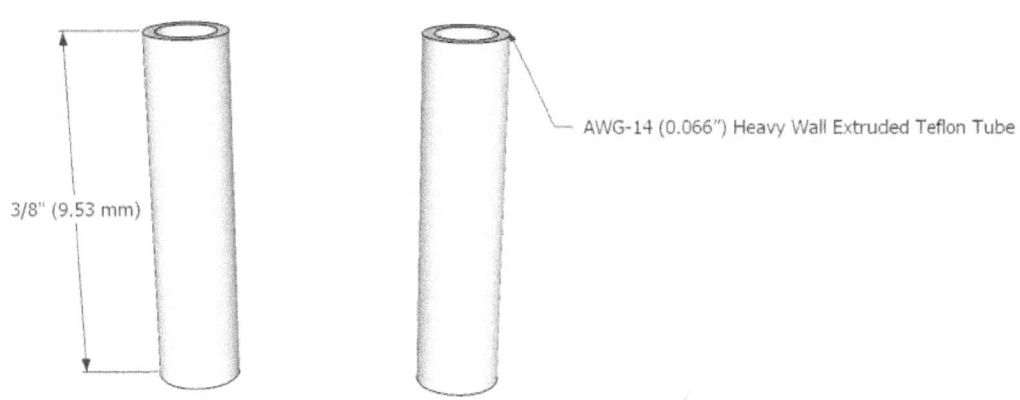

Part # 07 Displacer Crankshaft Bushing - AWG-14 (0.066" ID) Heavy Wall Extruded Teflon Tube, 3/8" (9.53 mm)
Part # 15 Drive Crankshaft Bushing - AWG-14 (0.066" ID) Heavy Wall Extruded Teflon Tube, 3/8" (9.53 mm)

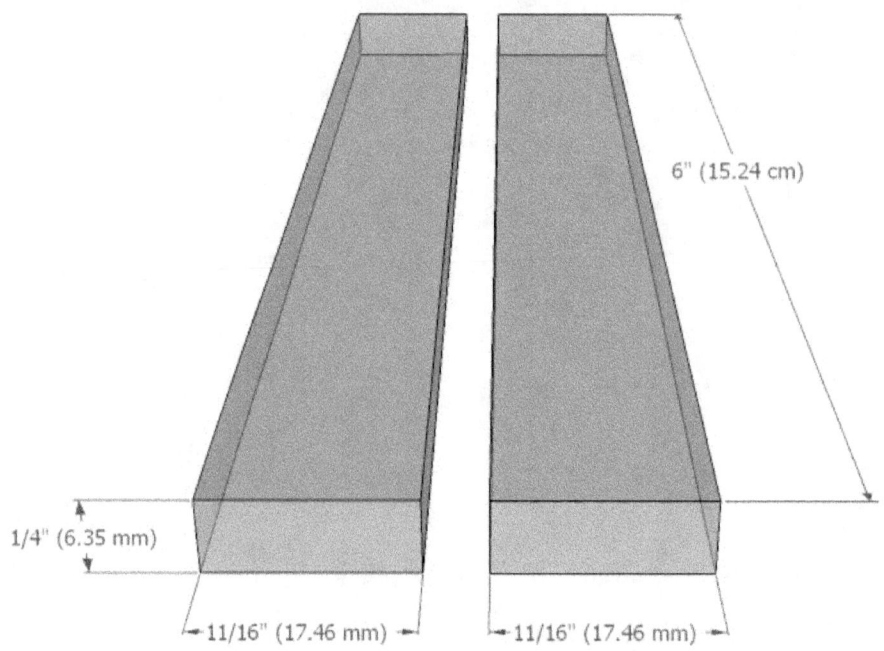

Part # 08 Long Sidewall - Clear Acrylic, 1/4" (6.35 mm) x 11/16" (17.46 mm) x 6" (15.24 cm) (2 pieces)

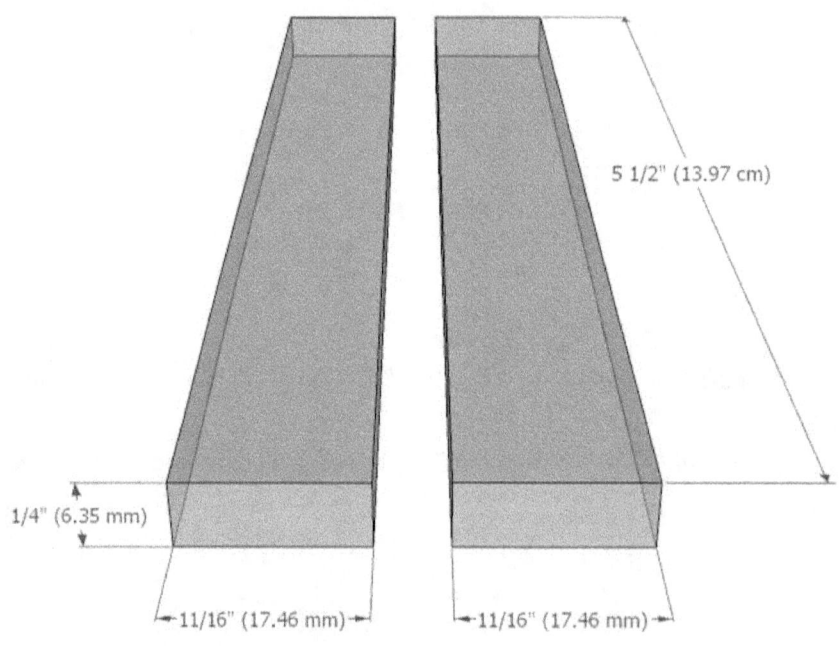

Part # 09 Short Sidewall - Clear Acrylic, 1/4" (6.35 mm) x 11/16" (17.46 mm) x 5 1/2" (13.97 cm) (2 pieces)

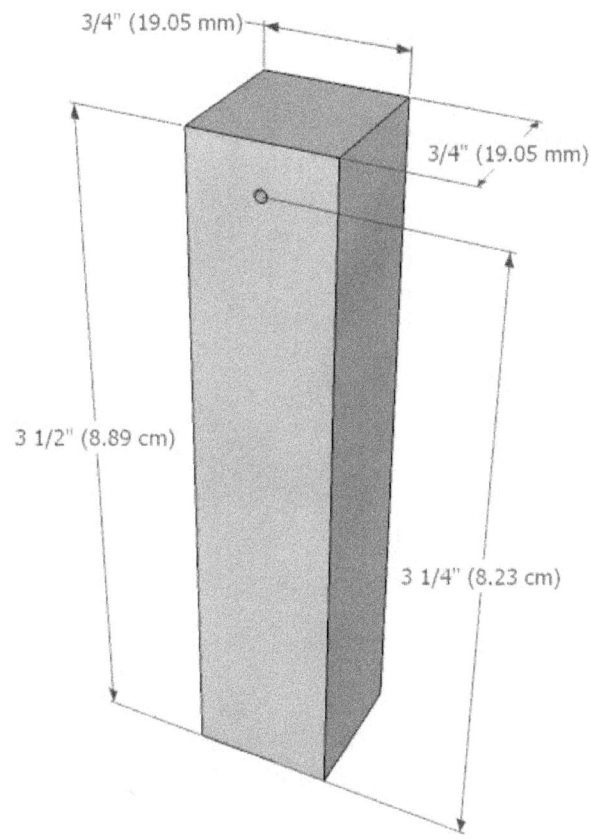

Part # 10 Pedestal - Wood, 3/4" (19.05 mm) x 3/4" (19.05 mm) x 3 1/2" (8.89 cm)

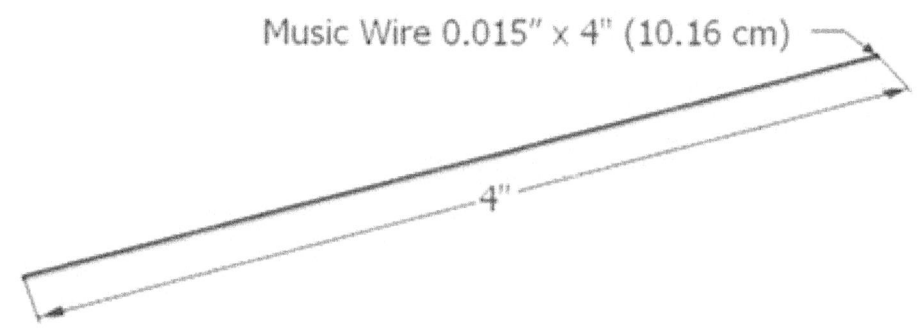

Part # 11 Displacer Connecting Rod - Music Wire, 0.015" x 4" (10.16 cm)
Part # 21 Drive Pushrod Music Wire, 0.015" x 4" (10.16 cm)
(2 Pieces)

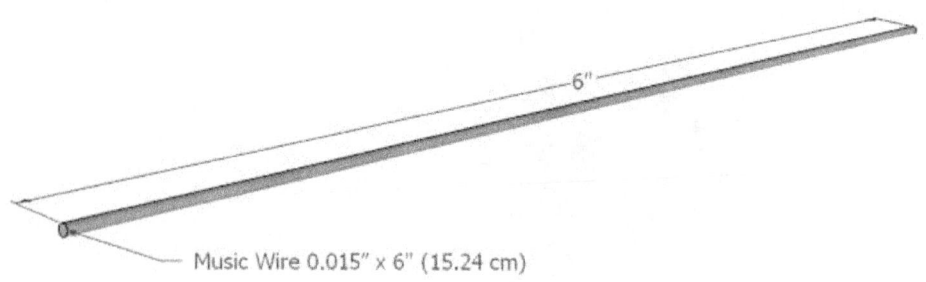

Music Wire 0.015" x 6" (15.24 cm)

Part # 12 Displacer Pushrod - Music Wire 0.015" x 6" (15.24 cm)

Duct Tape, 1/8" (3.18 mm) x 1/2" (12.7 mm) (2 pieces)

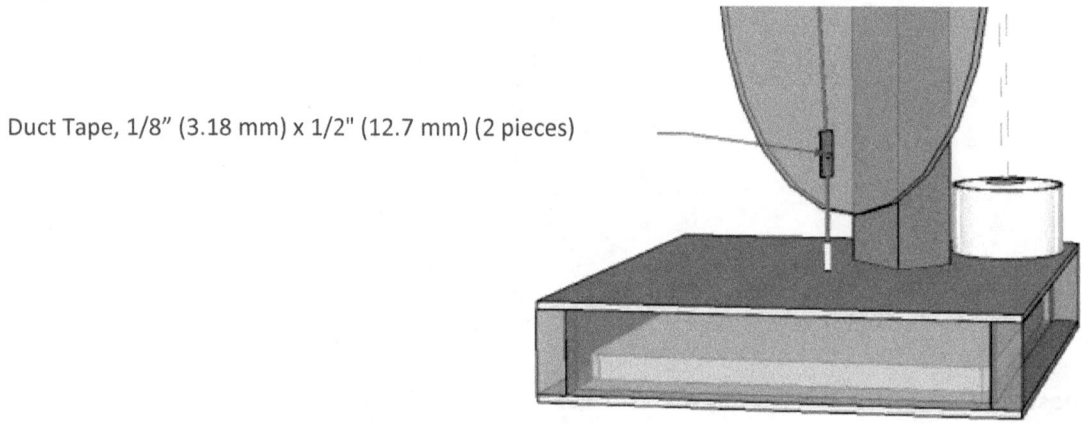

Part # 13 Displacer Pushrod Flex Joint - Duct Tape, 1/8" (3.18 mm) x 1/2" (12.7 mm) (2 pieces)

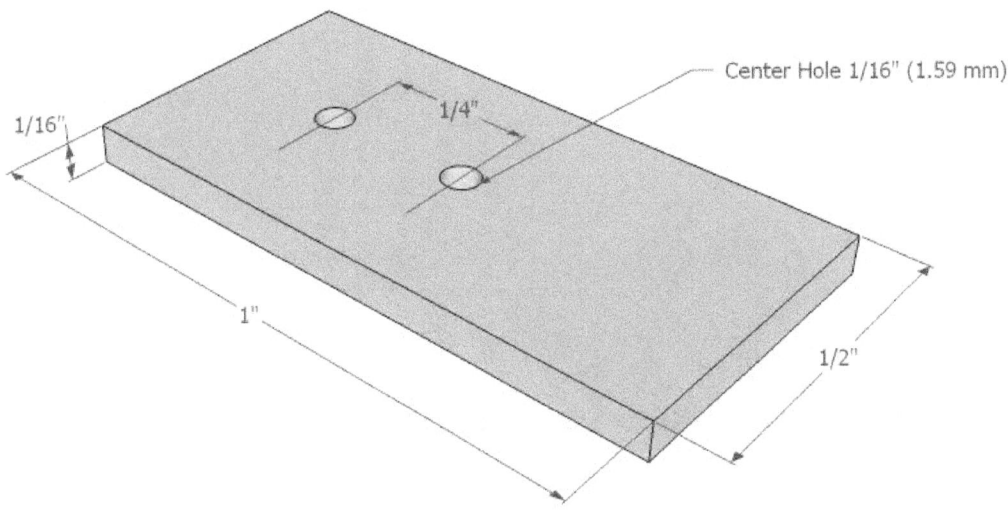

Part # 17 Drive Crankshaft Plate Aluminum Sheet, 0.062" (1/16", 1.59 mm), 1/2" (12.7 mm) x 1" (25.4 mm)

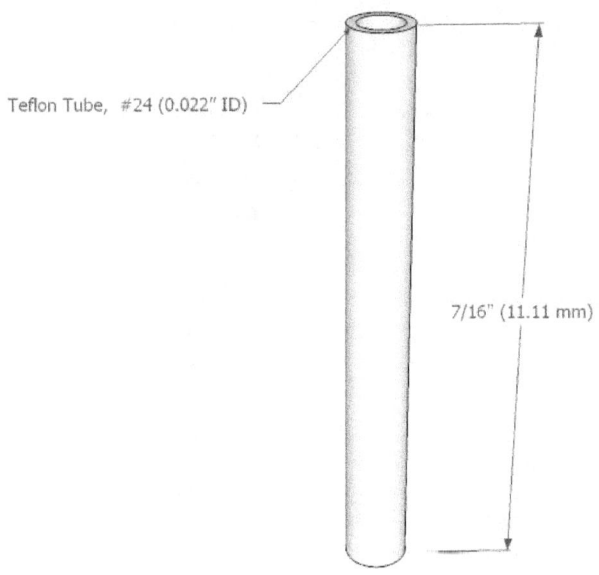

Part # 18 Gland - Teflon Tube, #24 (0.022" ID) x 7/16" (11.11 mm)

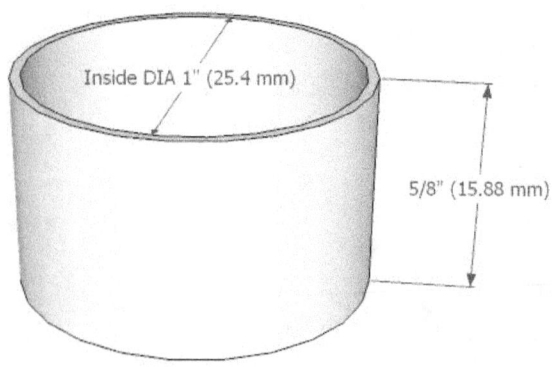

Part # 19 Drive Cylinder - PVC Pipe, 1" (25.4 mm) ID x 5/8" (15.88 mm)

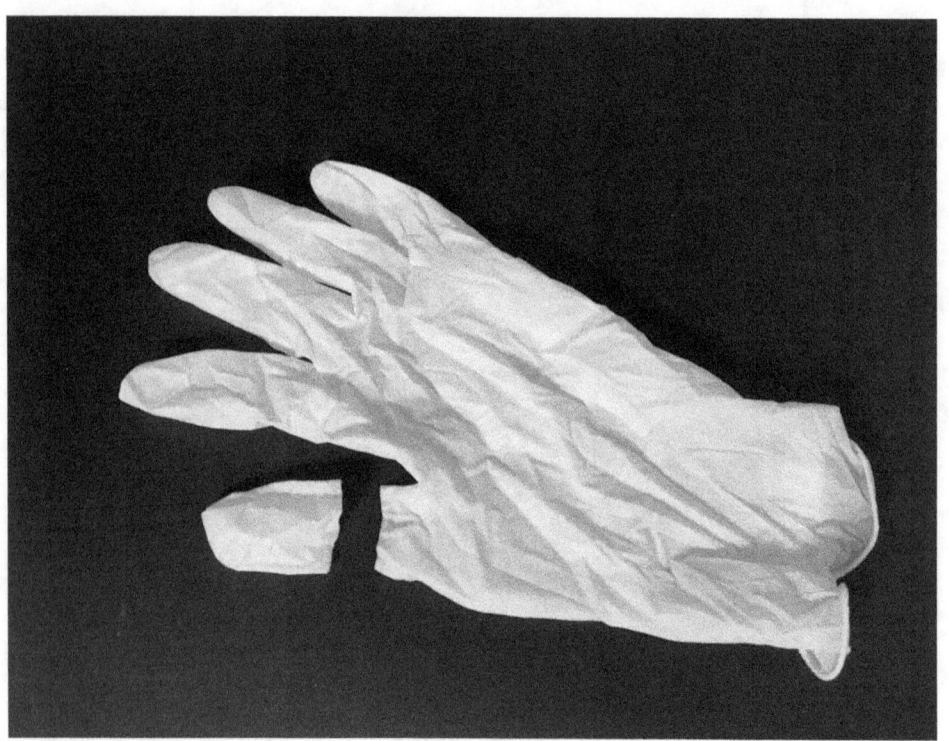

Part # 20 Drive Diaphragm - Latex glove fingertip

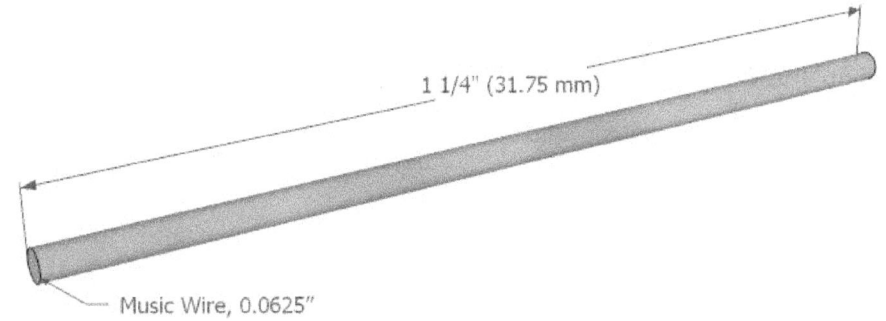

Part # 22 Axle - Music Wire, 0.0625"x 1 1/4" (31.75 mm)

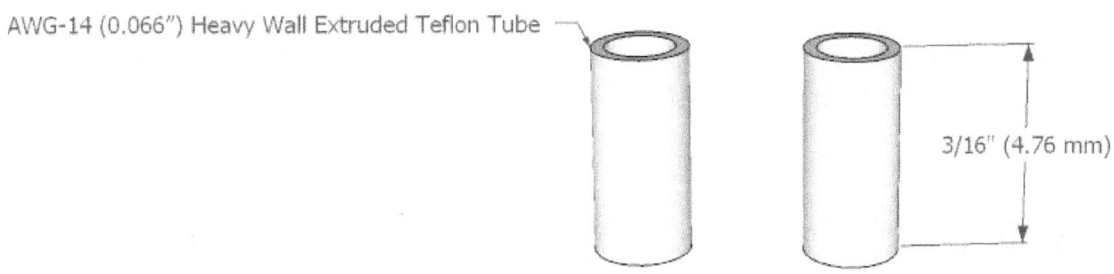

Part # 23 Axle Bushings - AWG-14 (0.066" ID) Heavy Wall Extruded Teflon Tube, 3/16" (4.76 mm) (2 pieces)

Assembly Instructions for Large Square Engine

Make the Pressure Chamber Top and Bottom Plates

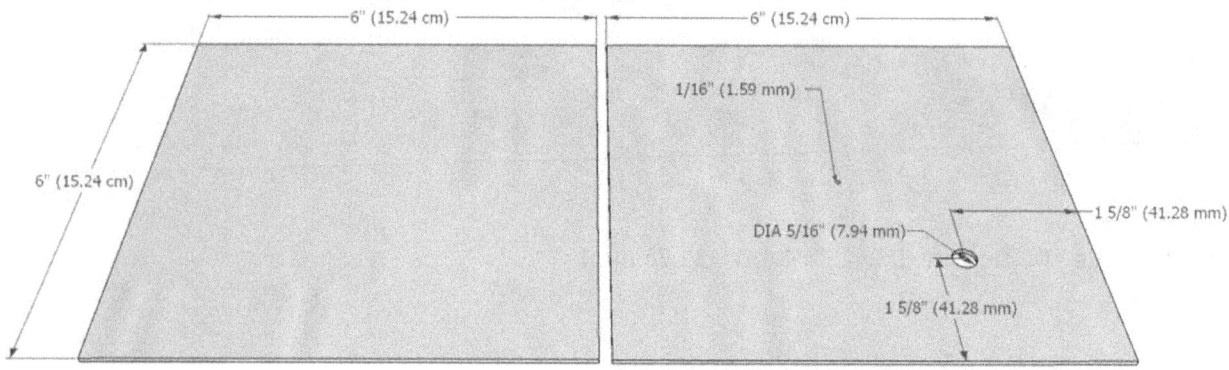

Cut two aluminum plates to form the top and bottom plates of the pressure chamber. Plate thickness is 0.0625" (1.59 mm). The plates are square, measuring 6" (15.24 cm) x 6" (15.24 cm).

Wash the metal parts with soap and water, rinse well and allow them to dry.

Paint all surfaces of the top and bottom plates (if desired). Use spray enamel that is suitable for painting metal.

Use a straight edge to find the exact center of the top plate. Use a center punch to mark the spot. Drill a hole that is the same diameter (or slightly larger) than the outside diameter of the Teflon tube that will be used for the displacer gland. The size will vary depending upon the material chosen.

Mark the position for the drive cylinder hole. The hole is located in one corner of the top plate, 1 5/8" (41.28 mm) from each side. The size of the hole is 5/16" (7.94 mm). Mark the hole position with a center punch and carefully drill the hole.

Make the Pressure Chamber Side Walls

Cut two short sidewalls and two long sidewalls.

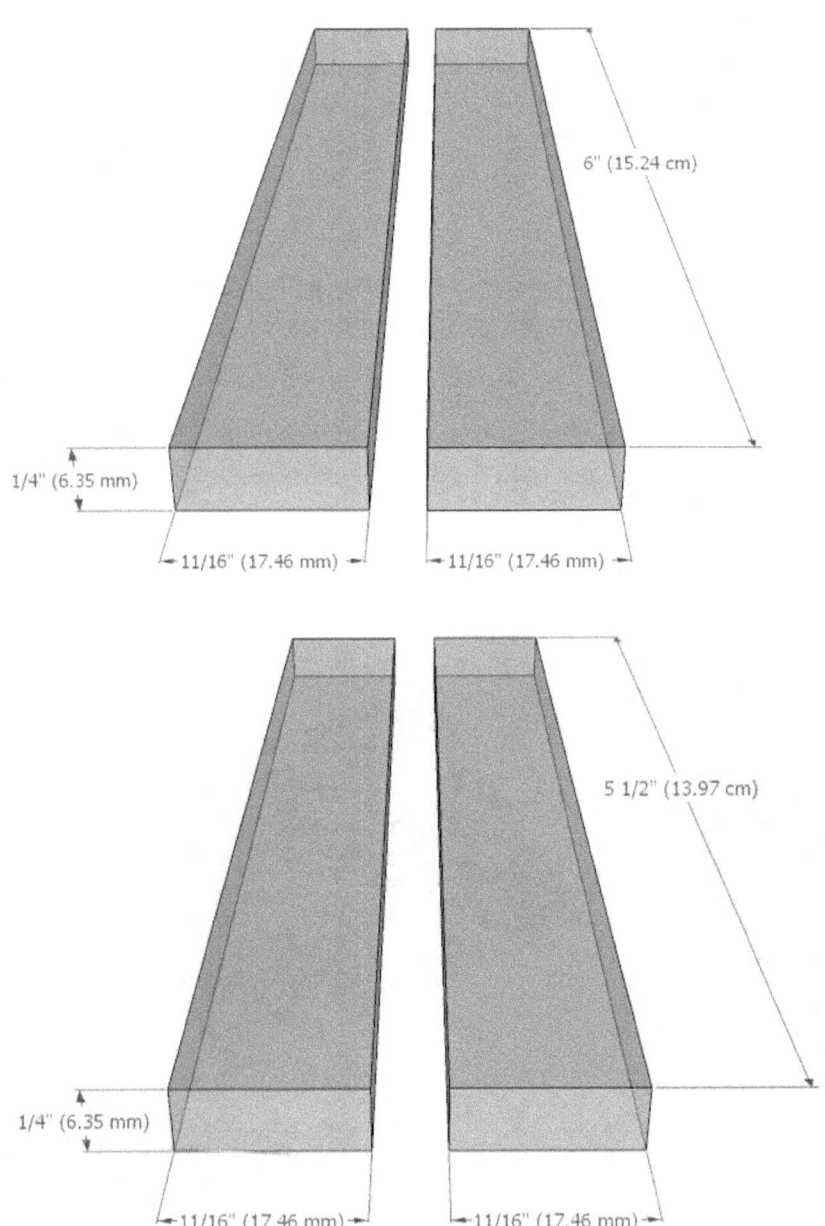

The sidewalls are cut from clear 1/4" (6.35 mm) acrylic sheet. The height of the sidewalls is 11/16" (17.46 mm). The length of the short sidewalls is 5 1/2" (13.97 cm). The length of the long sidewalls is 6" (15.24 cm).

It is also possible to make the sidewalls from slightly thinner 3/16" (4.76 mm) acrylic sheet.

Attach the Sidewalls to the Pressure Chamber Bottom Plate

Carefully check the fit of all the parts before assembly and use enough adhesive to ensure a good seal. Use clear five-minute epoxy to attach the acrylic sidewalls to the pressure chamber bottom plate. The sidewalls must be attached so that the pressure chamber is air-tight. The corner joints must also be sealed with glue. A small bead of glue oozing out of the joint indicates that the joint is being sealed well. Too much oozing glue, however, may interfere with the movement of the displacer inside the pressure chamber.

The top plate will _not_ be glued on at this time. The top plate may be used to help hold the sidewalls in place as the glue hardens.

Note: The glue will leave a permanent mark anywhere that it comes in contact with the acrylic. Be very careful not to create too many fingerprints on the acrylic during the gluing process, as these may be permanent.

Note: Silicone adhesive is another option for gluing the sidewalls to the pressure chamber. Silicone adhesives tolerate heat better than most clear epoxy glues, but they do not hold as well. Silicone is especially useful if there is a need to disassemble the engine for any reason.

The pressure chamber top plate, bottom plate, and sidewalls are ready for assembly.

Dry fit the parts before gluing and make any adjustment necessary to create a good fit.

Epoxy has been applied to attach the pressure chamber sidewalls to the bottom plate. The top plate is not being attached at this time. The top plate is holding the sidewall pieces in place as the glue cures.

Cut the Foam Displacer Panel

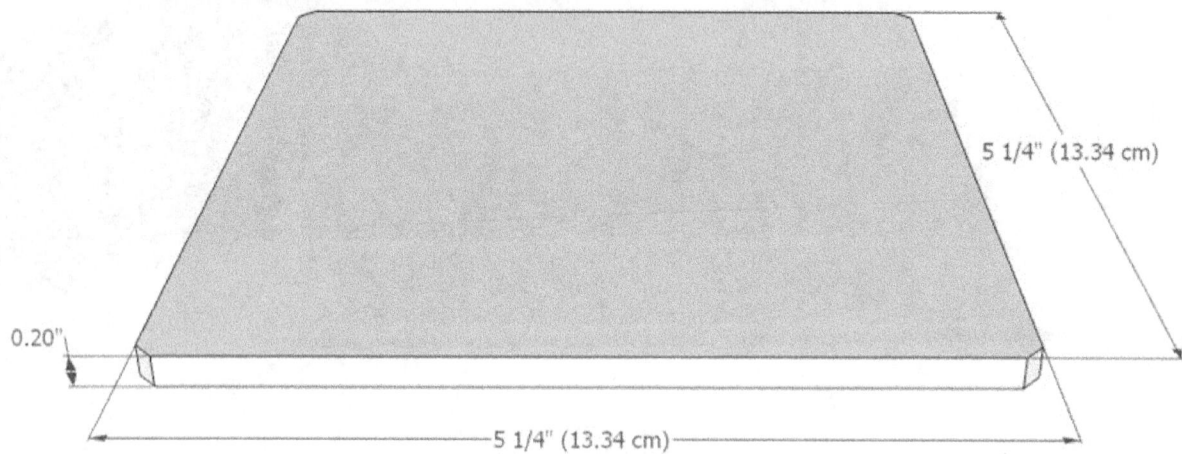

The displacer is made from Styrofoam that is cut to a thickness of 0.20" (13/64", 5.08 mm). For the best results, cut the foam with a hot wire foam cutter similar to the one described elsewhere in this book. The corners of the foam displacer have been rounded to prevent contact with the inside of the pressure chamber.

A hot wire foam cutter uses Nichrome wire that is heated by a small electrical current to cut the foam. The framework of the hot wire foam cutter holds tension on the wire to keep it straight, and a small electrical transformer provides the current to heat the wire. Foam cutters can be purchased from craft suppliers, or one can be built using simple parts. Plans for a homemade hot wire foam cutter are provided later in this book.

The dimensions of the displacer provide clearance between the displacer and the sidewall of the pressure chamber. That clearance needs to be 1/16" (1.59 mm) to 1/8" (3.18 mm) between each side of the displacer panel and the pressure chamber sidewall. Measure the internal dimensions of the pressure chamber and make any necessary adjustments to the size of the displacer before cutting.

Attach the Displacer Pushrod
Cut a piece of 0.015" music wire to a length of approximately 6" (15.24 cm). Small diameter music wire can be cut with ordinary wire cutters.

Make a 90° bend 1/2" from one end of the music wire.

A simple jig is used to hold the wire at the correct angle while attaching it to the foam displacer. Make the jig by drilling a small hole through a piece of flat wood. Drilling the hole in the jig will require a small diameter drill bit and a drill press. Choose a bit size that will provide a snug fit for the music wire.

If small drill bits under 1/16" (1.59 mm) are not available, use a 1/16" (1.59 mm) drill bit and insert a piece of Teflon tube into the hole to make a better fit for the music wire.

Use a straight edge to find and mark the exact center of the displacer foam. Pierce the foam with the long straight end of the displacer pushrod and press it in until the 90⁰ bend is pressed up against the surface of the foam. Place the protruding music wire into the hole in the jig and pull it though the jig until the foam is setting flush against the surface of the jig. The jig should now be holding the pushrod perpendicular to the foam displacer.

Press down on the wire at the center of the displacer to make a small depression in the foam no more than 1/32" (0.79 mm) deep. Fill the depression with high temperature epoxy to attach the pushrod to the displacer. Leave the displacer and pushrod in the jig until the epoxy has cured.

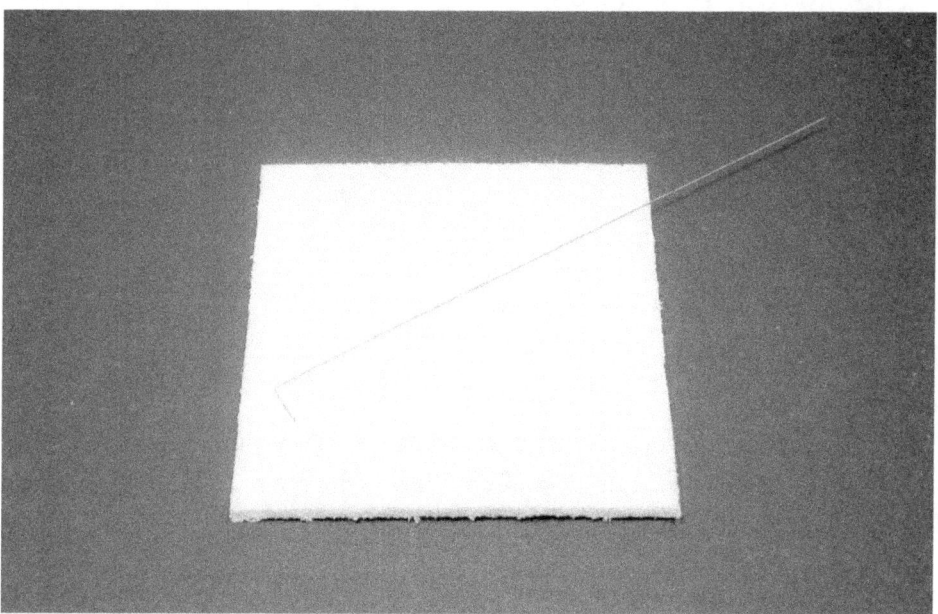

The displacer panel and the displacer pushrod are ready for assembly.

The displacer pushrod is held at the correct angle by the wooden jig. A hole is drilled through the board to hold the shaft at the correct angle. High temperature epoxy is used to attach the pushrod to the displacer panel.

Prepare the Drive Cylinder Tube

As noted earlier, the tubing used to make the drive cylinder is approximately 1" (25.4 mm) in diameter. Cut a piece of this tubing to a length of 5/8" (15.88 mm). Cut the pipe carefully to maintain a 90° angle on each end of the short drive cylinder.

Use fine sand paper to remove the glossy factory surface from the PVC pipe. This will also remove any labels and markings and improve the look of the finished part. Sanding the outer surface of the drive cylinder will help hold the latex drive diaphragm in place.

Place the drive cylinder over the hole near the corner of the pressure chamber top plate and glue it in place with epoxy. Apply the glue on the inside edge of the drive cylinder tube. Applying the glue on the inside edge will hide the glue joint from view when the engine is fully assembled.

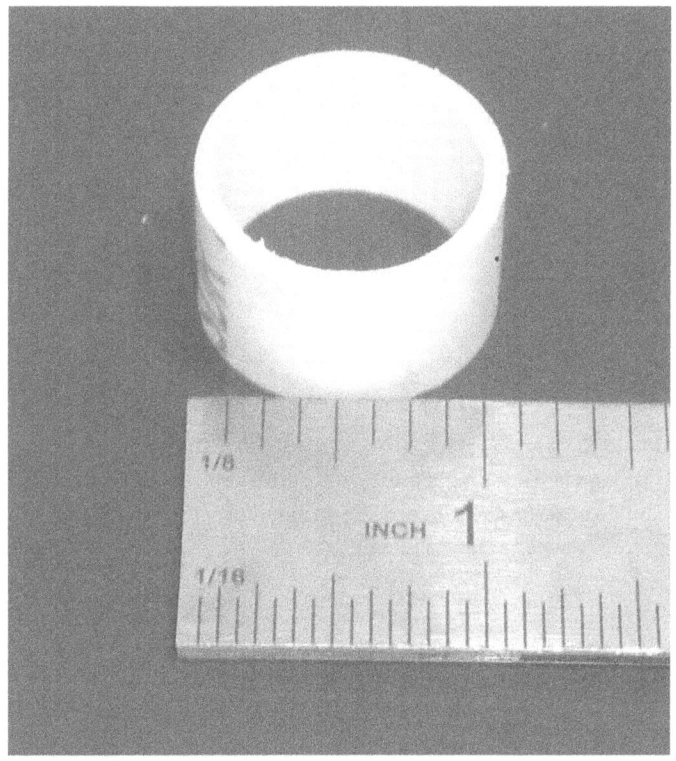

The drive cylinder is made from a short piece of 1" (25.4 mm) diameter PVC pipe.

The drive cylinder is attached with epoxy. The glue is spread on the inside of the cylinder to create a nice looking airtight joint.

The drive cylinder is glued in place on the pressure chamber top. The glue is on the inside of the cylinder to provide an airtight seal that will be hidden from view after final assembly.

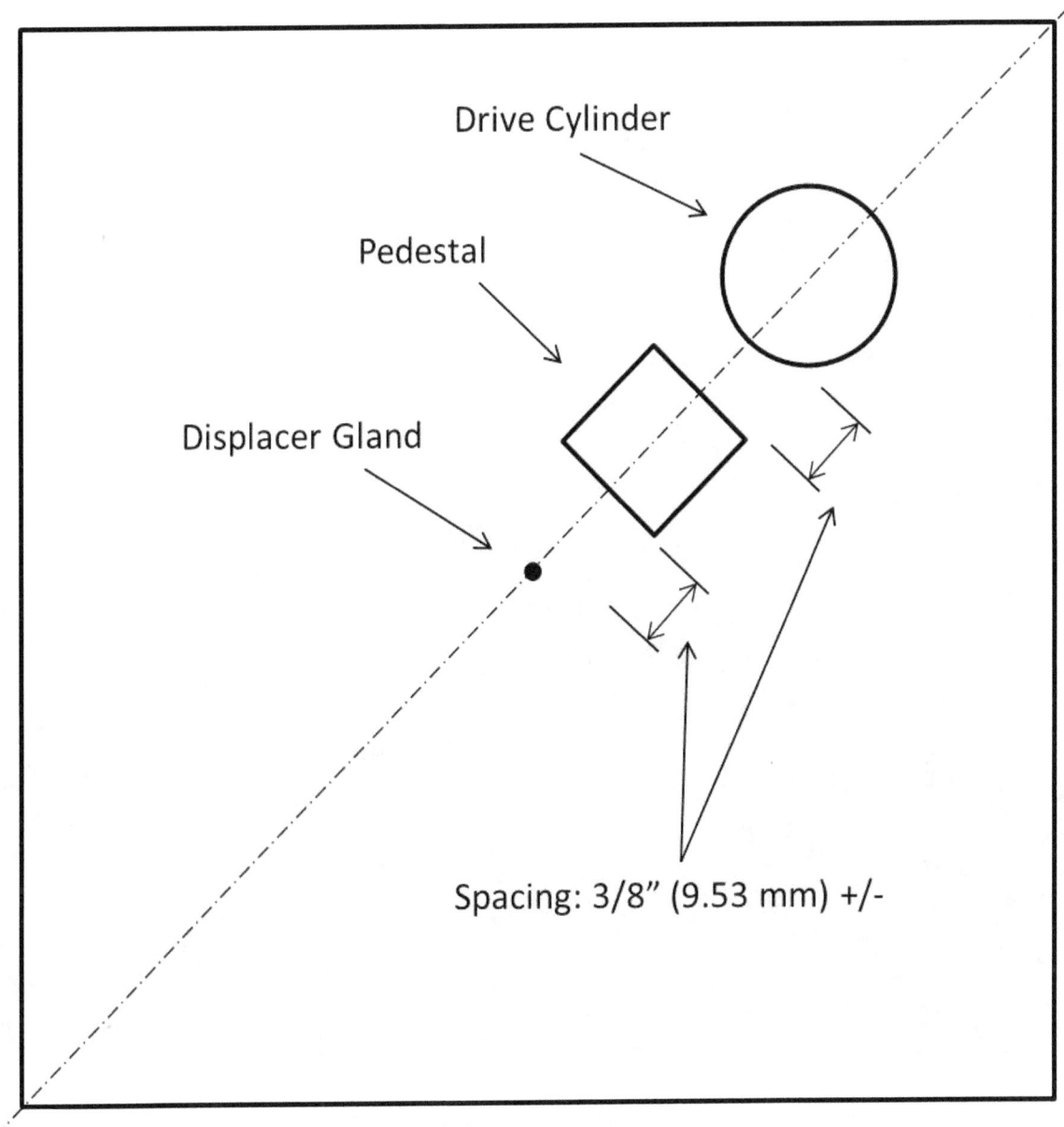

Drive Cylinder

Pedestal

Displacer Gland

Spacing: 3/8" (9.53 mm) +/-

This diagram shows the positioning of the parts on the top of the large square engine.

Attach the Displacer Gland

In the world of machinery, a "gland" is a sleeve within a stuffing box, fitted over a shaft in such a way as to prevent leakage of fluid while allowing a shaft or stem to move. Additionally, the gland of a Stirling engine allows the shaft to move with little or no friction. The gland is nearly air-tight. A tiny (very tiny!) pressure leak is desirable and helps the engine adjust for pressure changes as it warms up.

The gland is made of #24 (0.022" ID) Teflon tube. This provides a friction free fit for the 0.015" displacer pushrod.

Cut a piece of this Teflon tube to a length of 7/16" (11.11 mm).

Cutting Teflon tube requires a very sharp knife or razor blade, and a piece of 0.015" music wire. Insert the music wire into the Teflon tube before cutting. Place the tubing and wire on a hard flat surface. Press the blade of the knife against the tubing and roll the tubing on the flat surface until the knife has cut all the way around the tube. The wire in the middle of the tube will stop the knife from cutting all the way through the tube unless the tube is rolled. This will prevent the end of the tube from being crushed or deformed during the cutting process.

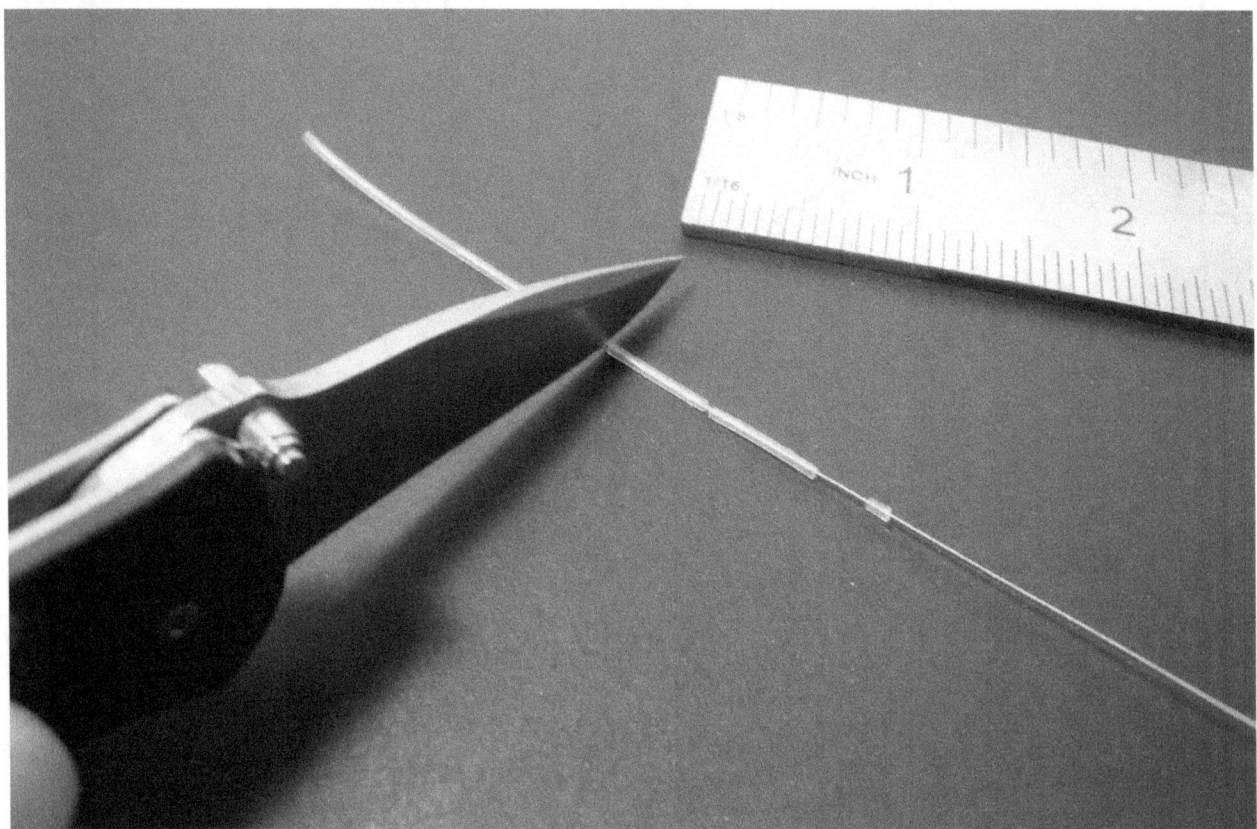

Insert a piece of music wire inside the Teflon tube before cutting. Roll the tube on a hard surface to cut around the wire. Cutting in this manner prevents the tube from deforming.

To attach the displacer gland:

1. Locate the displacer/pushrod assembled previously, the pressure chamber top plate, and the Teflon gland.
2. Slide the displacer pushrod through the top plate of the pressure chamber.
3. Set the displacer/top plate on a flat level surface with the pushrod pointing upward.
4. Slide the 7/16" (11.11 mm) gland tube onto the displacer pushrod.
5. Test for a friction free fit by verifying that the short piece of Teflon tube can fall under its own weight when dropped.

6. Let the Teflon gland drop inside the hole on the pressure chamber top plate until the end of the Teflon tube is flush with the inside surface of the top plate.
7. Apply a small bead of epoxy glue around the base of the gland tube, sealing it to the pressure chamber top plate. Take care that the glue does not touch the displacer pushrod. Allow the glue to cure.

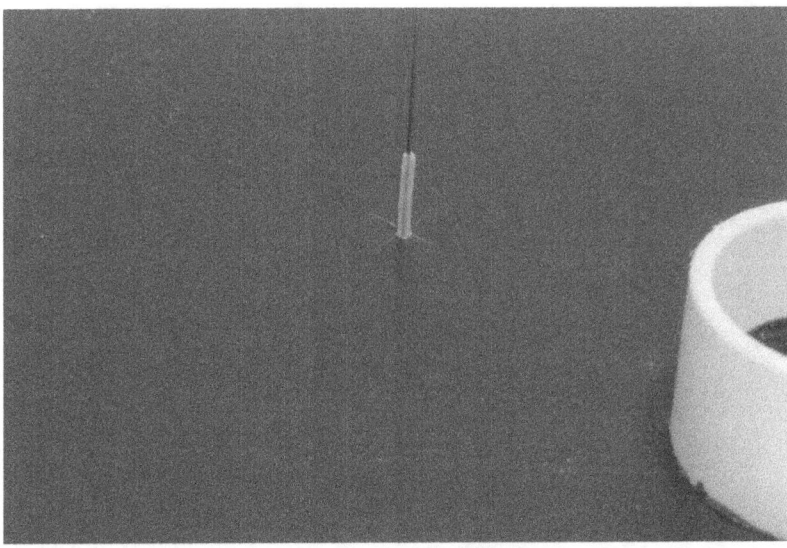

The pressure chamber top plate is placed over the displacer with the pushrod in place. The Teflon tube gland is shown here before glue is applied.

High temperature epoxy is applied around the base of the displacer gland.

99

Create the Flywheel

The flywheel is the largest and most visible moving part on this engine. If this part is built with care it will give the engine a very nice finished look. The goal is to make a flywheel that is perfectly round with a connection point for the axle that is squared to the flywheel and in the exact center. If the end result is less than perfect, the engine will still operate, and may even operate quite well. It just looks better if it is round, squared to the shaft, and centered.

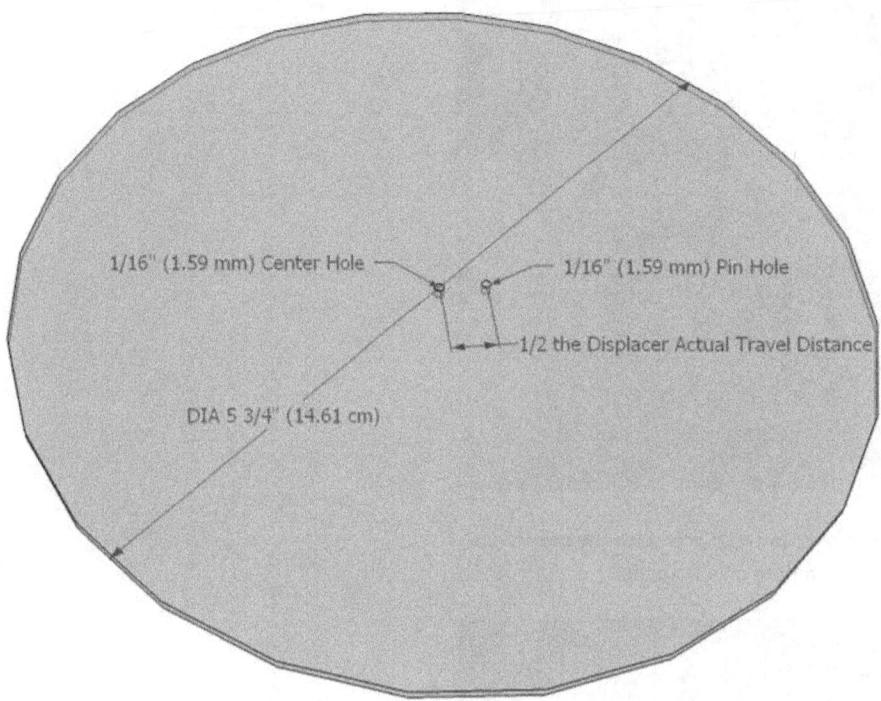

The flywheel for the large engine is 5 3/4" (14.61 cm) in diameter.

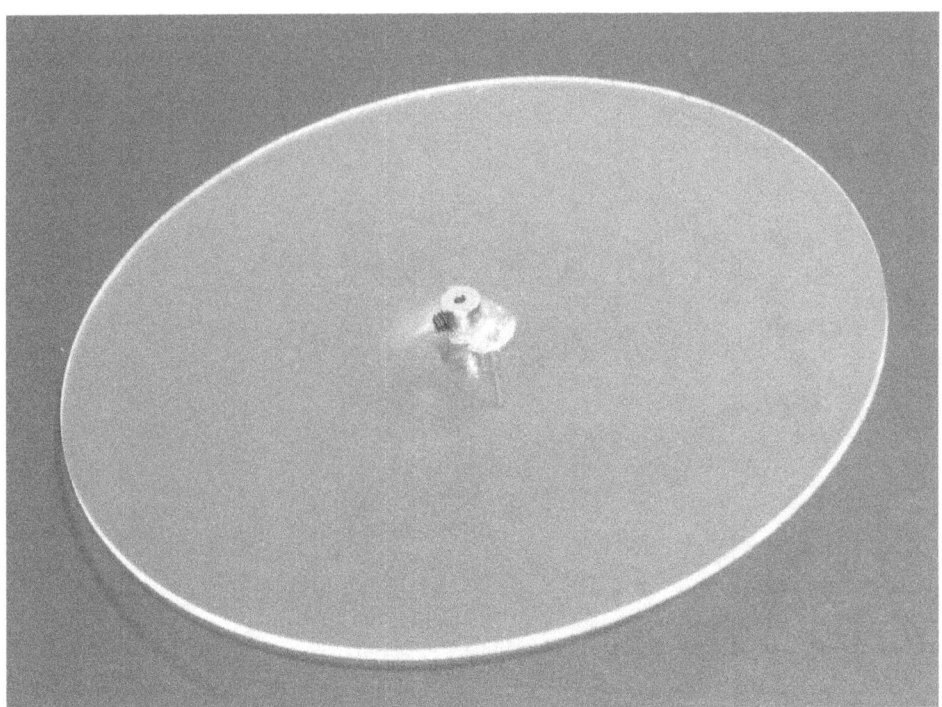

The flywheel pictured here was cut on a band saw using a simple homemade circle cutting jig.

Make a Simple Circle Cutting Jig

The process for creating a circle cutting jig and cutting flywheel is shown in detail in the instructions for the 4" Square engine in the previous chapter of this book. The process for making this flywheel is the same, with the exception that the diameter of this flywheel is larger. Please refer to the previous chapter for detailed instructions for how to make the flywheel.

Using a band saw and a circle cutting jig is the recommended method for making the flywheel. If a band saw is not available, the flywheel can be cut with hand tools.

Assemble the Pressure Chamber with the Displacer

Place the displacer pushrod through the gland and place the top plate on the pressure chamber with the displacer inside. Check the fit of all the parts and ensure that the displacer can move up and down without being obstructed. Make any necessary adjustments to the displacer before the top is glued onto the pressure chamber.

Use silicone adhesive to attach the pressure chamber top plate to the sidewalls. Spread a thin layer of the adhesive on the top edge of the sidewalls and smooth it with your finger. Use enough adhesive to create a seal, but no more. If too much adhesive is applied it may ooze into the inside of the pressure chamber and interfere with the motion of the displacer. Clamp the parts together with light pressure and set them aside until the silicone adhesive has cured.

Silicone adhesive is used for this joint because it can be opened up later if repairs are necessary.

Drill the Hole for the Crankshaft Pin in the Flywheel

The crankshaft pin is mounted in a hole that is drilled near the center hole of the flywheel. The spacing between the center flywheel hole and the crankshaft pin determines how far the displacer will travel as the flywheel rotates. The spacing between these two holes will be exactly half the _actual travel distance_ of the displacer.

Measure the travel of the displacer pushrod. To do this, take a measurement from the top of the pressure chamber to the top of the displacer pushrod when the displacer is on the bottom of the pressure chamber. Now lift up on the pushrod until the displacer is at the top of the pressure chamber and measure it again. Subtract the smaller number from the larger number to calculate the _total available travel distance_.

Subtract 1/16" (1.59 mm) from the _total available travel distance_ to get the _actual travel distance_. The actual travel distance will be slightly shorter than the total available distance so that the displacer will not touch the top or bottom of the pressure chamber as the engine runs. Shortening the travel distance by 1/16" (1.59 mm) will provide 1/32" (0.79 mm) of clearance above and below the displacer and prevent it from coming into contact with the pressure chamber.

It is important that the displacer does not come into contact with the pressure chamber as the engine is running. If the displacer hits the pressure chamber this will increase friction or prevent the engine from rotating freely. Also, it is good to keep a small cushion of air between the displacer and the pressure chamber top and bottom plates. The cushion of air reduces drag from what some refer to as "pull-off friction."

Pull-off friction can be demonstrated by holding a flat piece of cardboard against the ceiling with a broom handle. If the broom handle is quickly removed the cardboard does not immediately drop. The air pressure on the bottom of the cardboard holds it up until air is able to get in between the cardboard and the ceiling and equalize both pressures. This same effect can happen inside the pressure chamber if the displacer comes to rest in contact with the top or bottom plate of the pressure chamber. For this reason the engine is designed so that the displacer clearance is between 1/32" (0.79 mm) and 1/16" (1.59 mm).

Once you have determined the _actual travel distance_ of the displacer, divide that distance in half. This number will be the distance between the center hole of the flywheel and the hole for the crankshaft pin. Drill the hole for the crankshaft pin with a 1/16" (1.59 mm) drill.

The same circle cutting jig that was used to make the flywheel on the band saw is also an excellent jig for measuring and drilling the hole for the displacer crankshaft pin. The illustration shows how a pair of dividers can be used to measure the distance from the center pin to find the location for the displacer crankshaft pin hole.

Set the dividers to the distance needed for the offset of the crankshaft pin. Center one divider point on the center pin and the other point on the middle of the drill bit. Drill the hole when the alignment is correct.

Attach the Shaft Collar to the Flywheel

The flywheel is attached to the axle by means of a small round shaft collar that contains a set screw. The shaft collar is glued to one side of the flywheel in the exact center. Great care needs to be taken to align the shaft collar so that the main axle is perpendicular to the surface of the flywheel.

An alignment jig similar to what was used to attach the displacer pushrod will help attach the shaft collar with good alignment. Use a drill press to drill a 1/16" (1.59 mm) hole though a flat board. This board will serve as the alignment jig. Place a short piece of 1/16" (1.59 mm) music wire through the hole in the board. The music wire needs to be long enough to pass all the way through the board, the flywheel, and the shaft collar.

Place the flywheel over the music wire and press it flat against the surface of the board. Place the shaft collar over the music wire and press it flat against the flywheel. If the dry fit looks good, lift the shaft collar and spread a small drop of 5 minute epoxy under the shaft collar and press it against the surface of the flywheel.

The music wire will have to be removed before the epoxy is completely cured or it may become permanently attached. When the epoxy begins to harden, pull the music wire down through the alignment jig from the back side. This should leave the shaft collar glued over the center hole of the flywheel in near perfect perpendicular alignment.

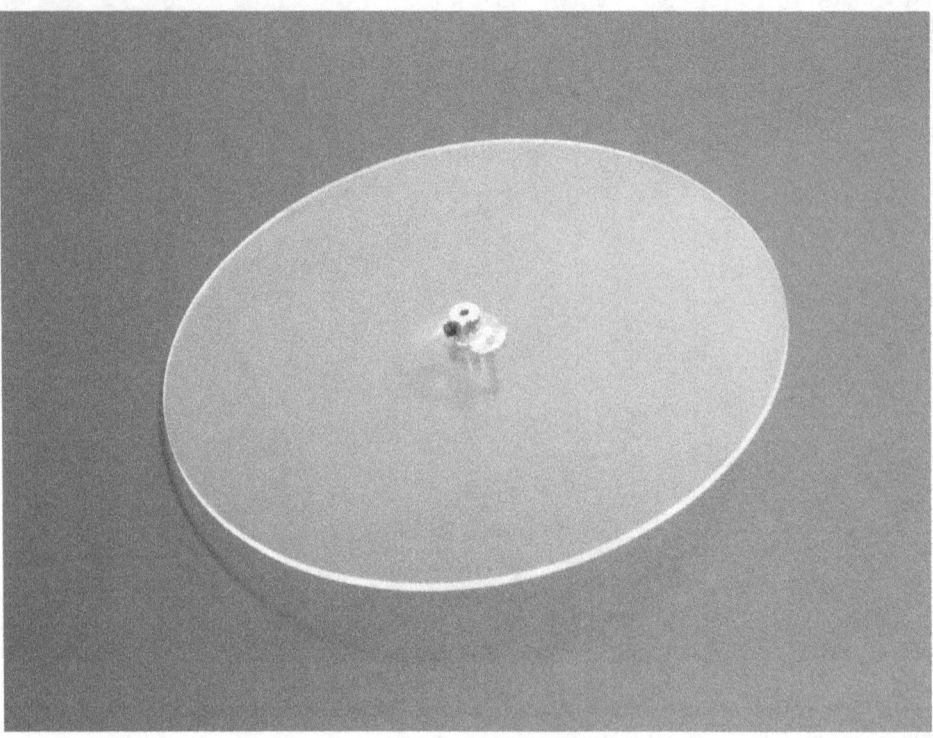

The shaft collar is attached over the center hole of the flywheel.

Make and Attach the Displacer Crankshaft Pin

There are two crankshaft pins. The displacer crankshaft pin attaches to a hole in the flywheel. The drive crankshaft pin is on the opposite end of the axle, over the drive diaphragm.

The displacer crankshaft pin is made from 0.0625" music wire, and is 1/2" (12.7 mm) long. The surface of the music wire should be smooth, with no gouges or tool marks, and it must be straight. Mark the wire where it is to be cut, clamp it in a vise, and then use the corner of a file to score the wire at the mark on two sides. Padding the vise and the pliers with paper will reduce the chances of scratching the music wire. Carefully bend the wire with pliers and it will break at the scored mark. Use a file or sandpaper to smooth the ends of the pin.

Glue the pin into the displacer pin hole on the flywheel. The location of the hole was calculated earlier. Use a small amount of epoxy to attach the pin to the flywheel. The pin must be perpendicular to the surface of the flywheel.

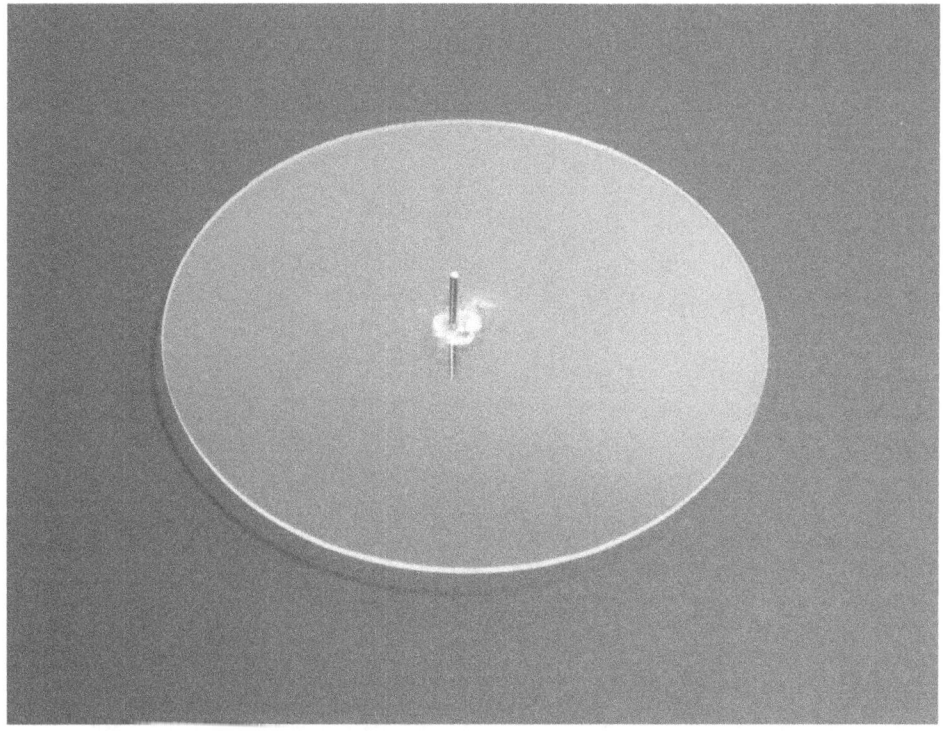

The crankshaft pin is attached to the flywheel on the opposite side as the shaft collar. The crankshaft pin is in the hole that is slightly off center.

Make the Main Axle

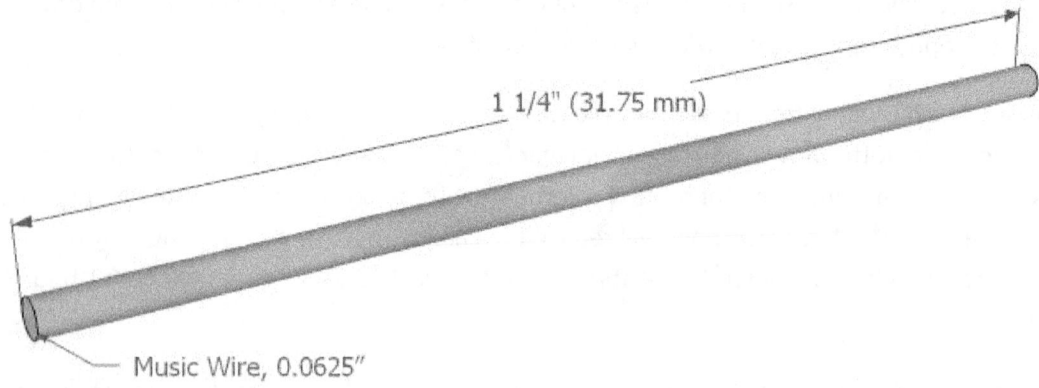

1 1/4" (31.75 mm)

Music Wire, 0.0625"

The axle is also made of 0.0625" music wire cut to a length of 1 1/4" (31.75 mm). Measure, score, and break the axle wire using the same technique that was used to cut the displacer crankshaft pin. Use a file or sandpaper to smooth the ends of the axle.

Make the Pedestal

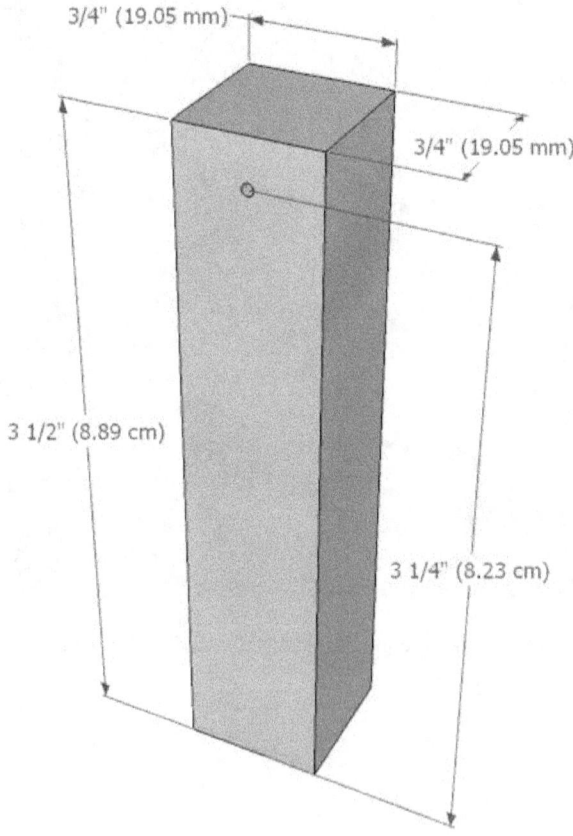

3/4" (19.05 mm)

3/4" (19.05 mm)

3 1/2" (8.89 cm)

3 1/4" (8.23 cm)

The pedestal is a small post that attaches to the top of the pressure chamber. It can be cut from a piece of wood. It measures 3/4" (19.05 mm) square by 3 1/2" (8.89 cm) tall.

Drill a hole in the top of the pedestal for the axle. Position the hole 3 1/4" (8.29 cm) from the bottom of the pedestal (which is 1/4" (6.35 mm) below the top of the pedestal). The size of the hole must provide a snug fit for the Teflon tube used for the axle bushings. The tubing used in this example fits snugly in a 7/64" (2.78 mm) hole. Measure to verify the hole size needed for the bushing material that will be used.

The pedestal may be finished with paint or varnish to improve the appearance of the finished engine.

Cut and Mount Teflon Tube Axle Bushings
The axle bushings are made from AWG-14 (0.066") Heavy Wall Extruded Teflon Tube. Two pieces are required. Each piece is 3/16" (4.76 mm) long.

Cut the tubing by first placing a piece of 0.0625" music wire inside the tube at the place to be cut. Place the tubing on a flat surface and roll the tubing as you cut it with a sharp knife or razor blade. Cut down to the music wire as the tubing is rolled on the flat surface. Cutting in this manner prevents the tubing from being deformed during the cutting process. Cutting the tubing without the shaft inside can cause the tubing to flatten or kink at the point of the cut, causing friction in the bushing.

Insert the small pieces of Teflon tube into each end of the hole in the pedestal. Leave a small amount of tubing (about 1/16" (1.59 mm) or less) protruding from the hole on both sides. No gluing will be required if the hole is the correct size.

The outside diameter of the axle is 0.0625", which allows it to turn freely inside the 0.066" ID Teflon tube. Insert a piece of axle material into the bushings to test the fit. Make any adjustments necessary to enable the axle to spin freely in the bushings.

The pedestal is shown here with the Teflon bushings ready to be installed.

Dry-Assemble and Measure for Locating the Pedestal

Attach the axle to the flywheel using the shaft collar, and insert this assembly into the bushings of the pedestal.

Position the pedestal on top the pressure chamber between the drive cylinder and the displacer pushrod. The axle must be centered over the displacer pushrod and the drive cylinder. The pedestal must be positioned so that both the pushrods can be attached to their respective pins at a 90° angle to the axle. Once the ideal position for the pedestal has been found, mark the position with a pencil. Set the pedestal aside for now. It will be attached after the drive crankshaft has been assembled.

Make the Drive Crankshaft

The drive crankshaft plate creates the offset for the crankshaft that attaches to the drive diaphragm. It holds the drive crankshaft pin parallel to the axle with an offset of 1/4" (6.35 mm). A hole is drilled to attach the crankshaft pin to one side. A shaft collar is glued to the opposite side and is the attachment point for the main axle.

Draw the shape of the crankshaft plate on a piece of aluminum stock and drill the hole before cutting the plate. The plate is made from 0.062" aluminum sheet, the same material recommended for the top and bottom plates of the pressure chamber. The dimensions of the plate are 1/2" (12.7 mm) x 1" (25.4 mm). Drill a 1/16" (1.59 mm) hole at a position 1/4" (6.35 mm) from one center of the plate. Cut the small plate from the aluminum sheet after the hole has been drilled.

Cut the drive crankshaft pin from a piece of 0.062" music wire. The length of the pin is 1/2" (12.7 mm). Attach the pin to the hole in the plate with epoxy. The pin must be perpendicular to the surface of the plate. The pin is mounted on the front side of the plate.

Place a mark on the back side of the plate that is 1/4" (6.35 mm) away from the center of the pin. This mark will be used to position the shaft collar to the back side of the plate. Use epoxy to attach a 1/16" (1.59 mm) shaft collar to the back side of the plate. Take care that the set screw of the shaft collar does not become fouled with epoxy.

The pictures show multiple crankshaft plates. Only one is required. Making several crankshaft plates with different offset measurements makes it possible to change the travel distance of the drive mechanism. This can be a useful adjustment for fine tuning the engine to operate in different temperature environments.

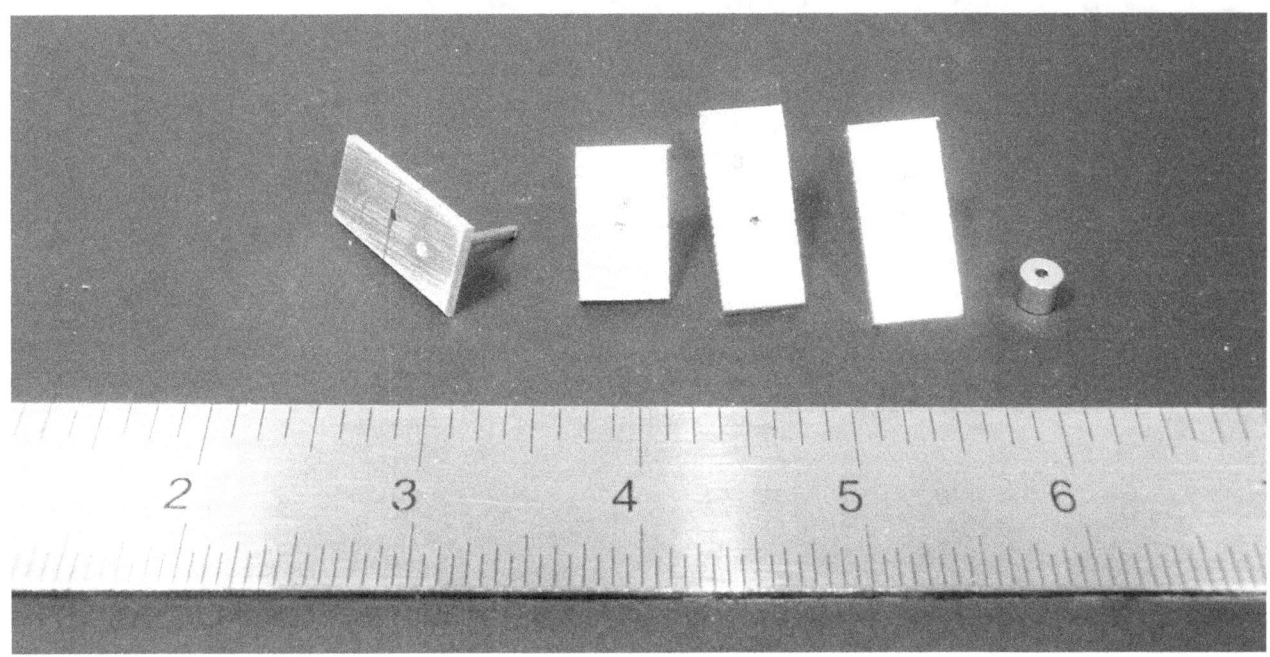

The crankshaft plate and the shaft collar are ready to be joined together with epoxy.

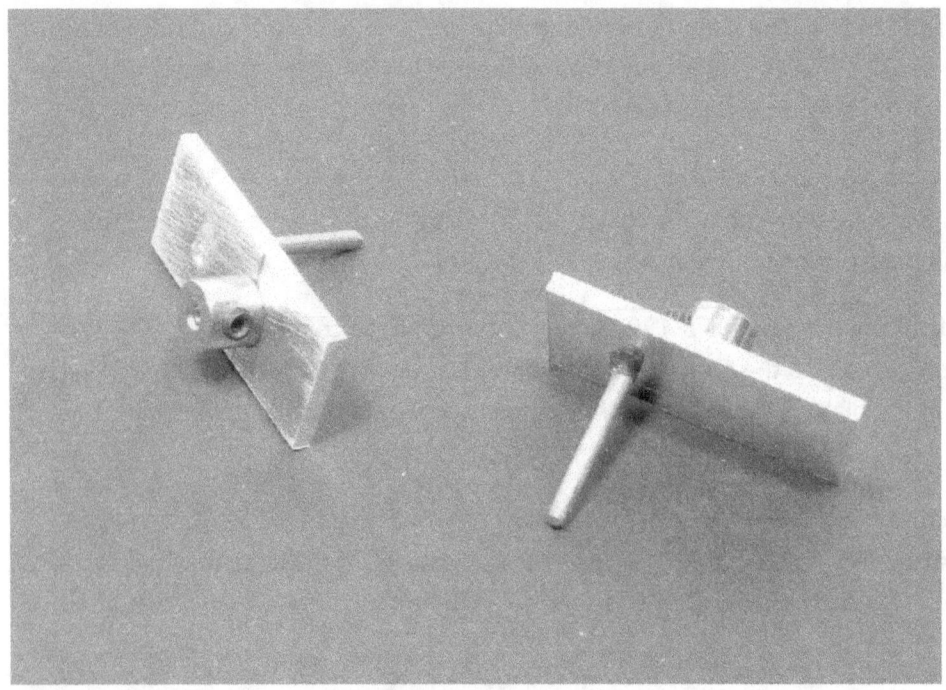

The crankshaft pin, plate, and shaft collar have been assembled.

Attach the Pedestal to the Pressure Chamber Top

The pedestal needs to be mounted so that it is aligned well with the displacer pushrod and the drive cylinder. Place a piece of straight music wire through the axle bushings to help align the pedestal. This will make is easier to visually check the alignment and confirm the previous marks. Glue the pedestal in place with epoxy once the correct position is confirmed.

The pedestal is attached to the pressure chamber top. There is enough room for the flywheel between the pedestal and the displacer pushrod, and the drive crankshaft pin extends over the middle of the drive cylinder.

Test the Travel Distance of the Displacer

Install the axle, flywheel, and drive crankshaft on the pedestal. Position the flywheel so that the displacer crank pin is at the bottom position of its rotation. Use a marker or a piece of tape to make a reference mark on the displacer pushrod at the point where it comes in contact with the crank pin. Now rotate the flywheel until the pin is at the top of its rotation and raise the displacer pushrod until the mark is once again even with the pin. There should be about 1/16" (1.59 mm) of free space between the displacer panel and the top of the pressure chamber when the displacer crank pin is at the top of its rotation. If it appears that the

111

displacer will be able to move up and down without impacting the top or bottom of the pressure chamber, proceed to the next step. If the displacer crank is moving the displacer too far and it is impacting the engine, correct the problem by relocating the crank pin closer to the center of the flywheel.

Trim the Displacer Pushrod
Allow the displacer panel to rest on the bottom of the pressure chamber. Measure up from the top of the pressure chamber 1 1/2" (38.1 mm) and trim the displacer pushrod at this point.

Create the Displacer Connecting Rod and Teflon Bushing
The displacer connecting rod completes the connection between the top of the displacer pushrod and the displacer crank pin on the flywheel. A flexible connection is made to the displacer pushrod with two thin pieces of duct tape. The top end of the connecting rod is a piece of Teflon tube that will slip over the displacer crank pin on the flywheel. The displacer connecting rod is made from 0.015" music wire, which is the same size as the displacer pushrod.

Cut a piece of AWG-14 heavy wall extruded Teflon tube to a length of 7/16" (11.11 mm). Use the same cutting method described earlier so that the tubing does not become deformed at the cut.

Cut a piece of 0.015" music wire to at least 3 1/2" (88.9 mm) long to construct the connecting rod. The exact length is not critical because it will be trimmed to fit. It may be easier to work with a longer piece and trim it to length after the bends are completed.

Place the short piece of Teflon tubing on a piece of axle stock (0.062" music wire) to help keep it straight while wrapping the connecting rod wire around the tubing. Make at least 1 1/2 wraps around the tubing with the connecting rod wire. Adjust the connecting rod wire so that it is at a 90° angle to the Teflon tube. The connecting rod wire and the Teflon tube should form the shape of a "T." Make the wraps tight enough so that they grip the Teflon tube and it does not fall out. Trim any excess wire when finished.

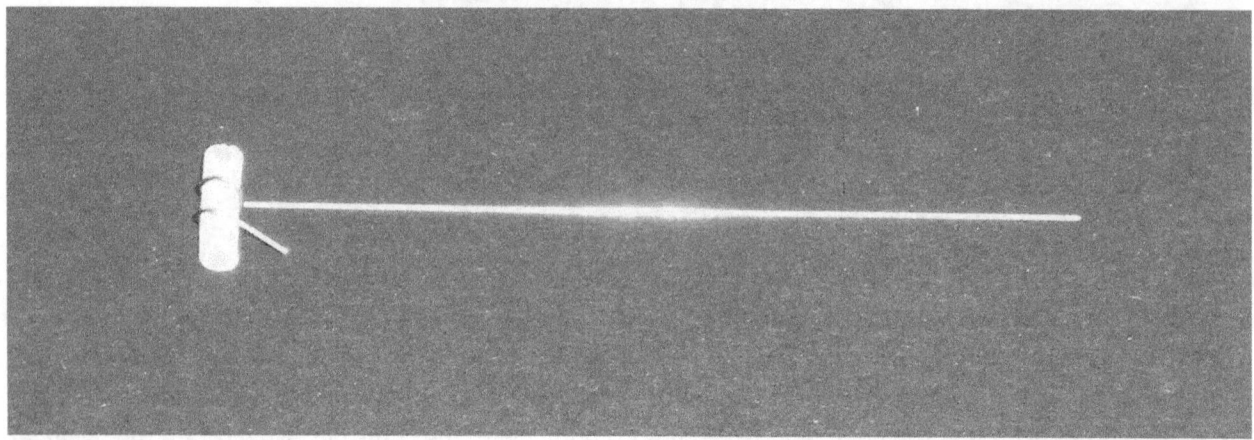

The Displacer Connecting Rod is fashioned from a length of music wire that is wrapped around a short piece of Teflon tube.

Attach the Displacer Connecting Rod

The connecting rod can now be installed between the flywheel and the displacer pushrod. Place the Teflon tube over the drive crankshaft pin on the flywheel. With the displacer panel resting on the bottom of the pressure chamber and the flywheel pin in its lowest position, trim the length of the connecting rod so that there is a gap between the end of the displacer pushrod and the end of the connecting rod. The gap should be between 1/32" (0.79 mm) and 1/16" (1.59 mm).

Cut two small pieces of duct tape to 1/8" (3.18 mm) x 1/2" (12.7 mm). Lift the displacer so that the end of the displacer pushrod is near the end of the connecting rod and fasten the two pieces together with the small pieces of duct tape. Position the duct tape so that the joint bends correctly in order to accommodate the motion of the flywheel.

Rotate the flywheel and observe the motion of the displacer panel. It should travel up and down inside the pressure chamber without touching the top or the bottom of the pressure chamber. Make any adjustments necessary so that the displacer does not come into contact with the top or bottom of the pressure chamber.

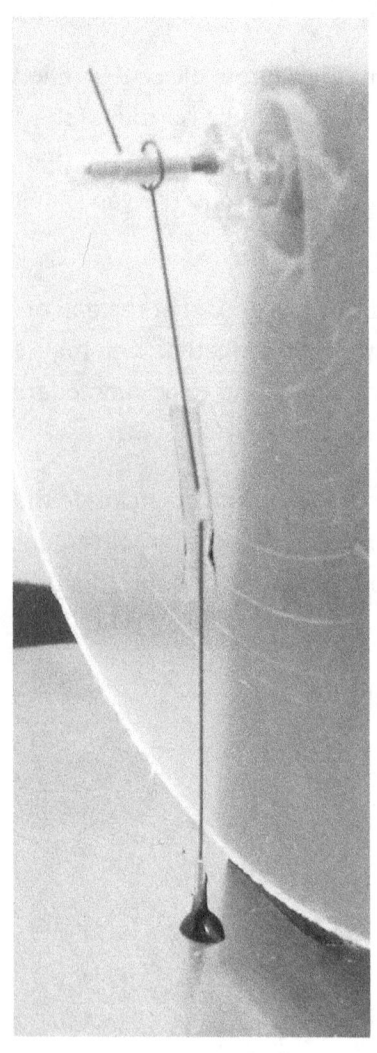

In this picture, the displacer connecting rod has been trimmed to leave a small gap between the ends of the two rods. One piece of duct tape has been applied to the joint. Note the direction of the bend at the joint and how the alignment of the tape allows it to act as a hinge.

Create the Drive Pushrod and Teflon Bushing

The drive pushrod is made just like the displacer connecting rod was made, except there is no duct tape joint in the middle of the shaft. There is a Teflon tube at the top end of the pushrod. The Teflon tube rides on the crankshaft pin. The bottom end is bent into a loop that is folded over so that it mounts flat against the drive diaphragm.

Cut a piece of AWG-14 Teflon tube to a length of 7/16" (11.11 mm). Insert a piece of 0.062" music wire inside to help hold it while wrapping a piece of 0.015" music wire around it. Make at least 1 1/2 turns around the tubing, as before. Adjust the tubing so that it is held snugly at a 90° angle to the pushrod.

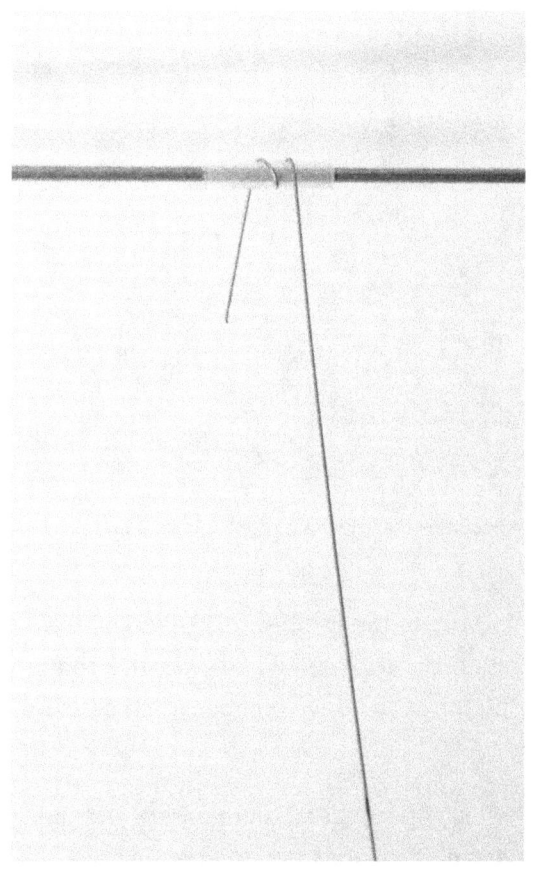

In this picture, the Teflon tube has been placed on a piece of axle wire and the pushrod wire is wrapped around the tubing to attach it to the pushrod.

Measure the distance between the center of the axle and the top of the drive cylinder. This will be the finished length of the pushrod.

Make a 90° bend in the pushrod using the measurement just obtained. The distance from the axle to the top of the drive cylinder should be the same as the distance from the Teflon tube to the 90° bend.

Use needle nose pliers to make a loop at the bottom of the pushrod. The loop should be about 7/16" (11.11 mm) in diameter. The loop is made in such a way so that if the connecting rod was placed on a flat surface it could stand upright with the loop flat against the table top.

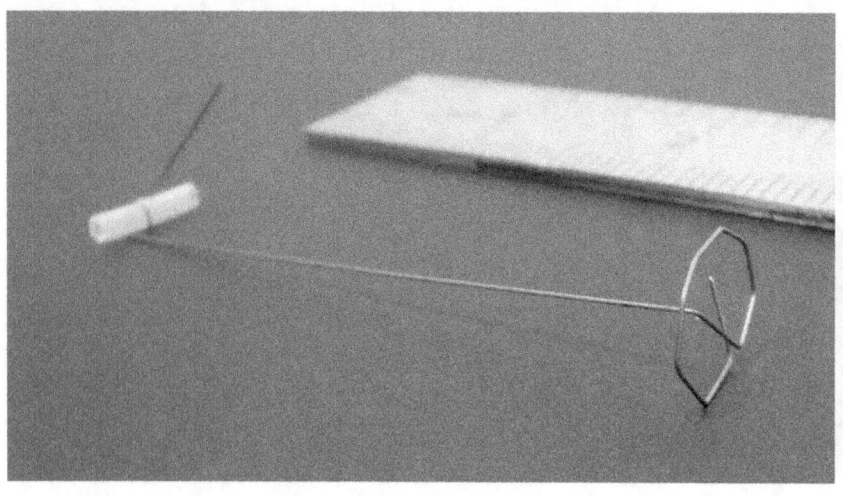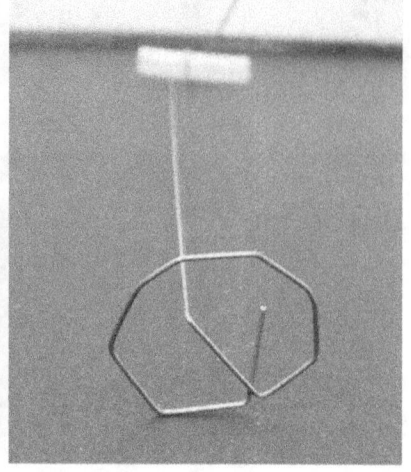

These pictures show the loop formed at the 90° bend in the pushrod. The loop becomes the attachment point for the drive diaphragm.

Verify the fit of the pushrod. Place it on the drive crankshaft pin and rotate the flywheel. The loop on the bottom should rise above the top of the drive cylinder at its highest point, and it should drop down into the drive cylinder at its lowest point. It should be close to even with the top of the drive cylinder when at its midpoint between high and low.

If getting the correct length is a challenge, consider making the shaft slightly longer than required, and then place a Z shaped bend near the middle of the pushrod. The Z bend can be manipulated to fine tune the length of the pushrod after it is installed.

Create the Drive Diaphragm from a Latex Glove

The drive diaphragm is made from the fingertip of a latex glove. Wash and dry a latex glove so that all the powder is removed. Cut the fingers from the glove. Stretch one of the glove fingers over the top of the drive cylinder. Pull the latex down the outside of the drive cylinder until there is only a small amount of slack near the center of the drive diaphragm.

The drive diaphragm should stay in place without any help. If it appears to be slipping, secure it by placing a rubber band around the outside of the drive cylinder. A rubber band can be made from another glove finger if necessary.

Attach the Drive Pushrod to the Drive Diaphragm

The loop at the end of the drive pushrod should rest flat against the drive diaphragm and there should be no sharp wires threatening to puncture the diaphragm. Hold the loop against the center of the diaphragm and attach it using Superglue®.

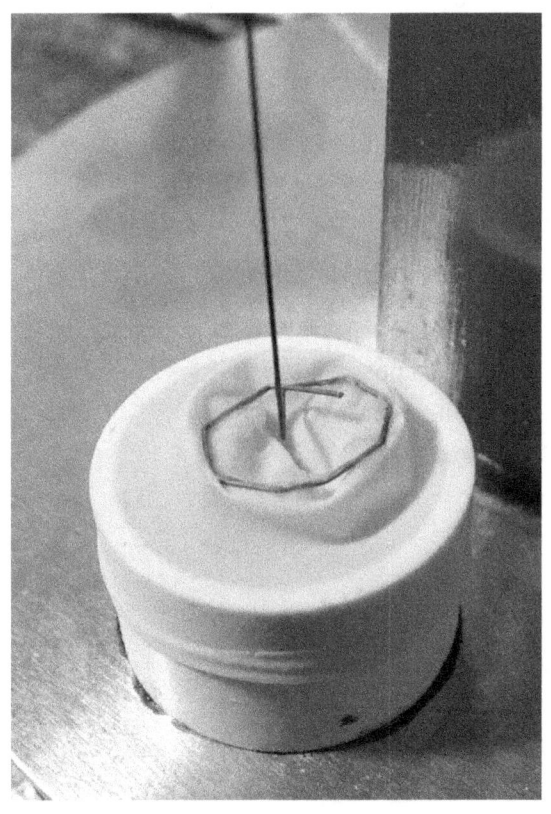

This picture shows the drive pushrod attachment to the drive diaphragm. The wire pushrod is attached to the latex diaphragm with Superglue®. The diaphragm is made from the finger tip that has been cut from a latex glove.

Set the Crankshaft Timing Angle

Adjust the flywheel and the drive crankshaft so that there is a 90° offset between the displacer crank pin and the drive crank pin. The direction of the offset will determine the direction the motor rotates when running. The drive mechanism will follow 90° behind the motion of the displacer mechanism. That means when the displacer is all the way up at the top of the rotation, the drive mechanism will be halfway up. When the displacer is all the way down, the drive mechanism will be halfway down. Use the set screws on the shaft collars to hold the flywheel and the drive crankshaft in place.

Adjust the Drive Diaphragm Tension

The drive diaphragm should be adjusted so that there is just enough slack in the latex to allow the engine to rotate without stretching the material. If the material is too tight the engine will not run well because extra energy will be required to stretch the diaphragm. If the diaphragm is too loose the engine will not run well because the loose diaphragm will inflate and deflate without causing the crankshaft to move. Adjusting the tension of the drive diaphragm is one of the adjustments that can be made to fine tune the performance of the motor.

Check all the Connections

The engine should now be fully assembled and ready for its first run. Check all the connections by rotating the flywheel slowly and observing all the moving parts. Nothing should be falling apart when the flywheel is rotated. If the connecting rod or the pushrod becomes disconnected during this test, make adjustments so that they do not fall off.

Observe the motion of the displacer panel inside the pressure chamber and ensure that it does not impact the top or the bottom of the pressure chamber. It should move without any obstruction.

Run the Engine!

This Stirling Engine should run well over hot water. Fill a coffee cup or similar container with near-boiling water. Place the pressure chamber on top the cup of hot water. Allow it to warm up for 10 to 20 seconds. Turn the flywheel to start the engine.

The motor will continue to run as long as there is a temperature differential of 20^0 F (11^0 C) (or more) between the top and bottom surfaces of the pressure chamber. It may be possible to fine-tune the engine to operate on an even lower temperature differential with a little care and patience.

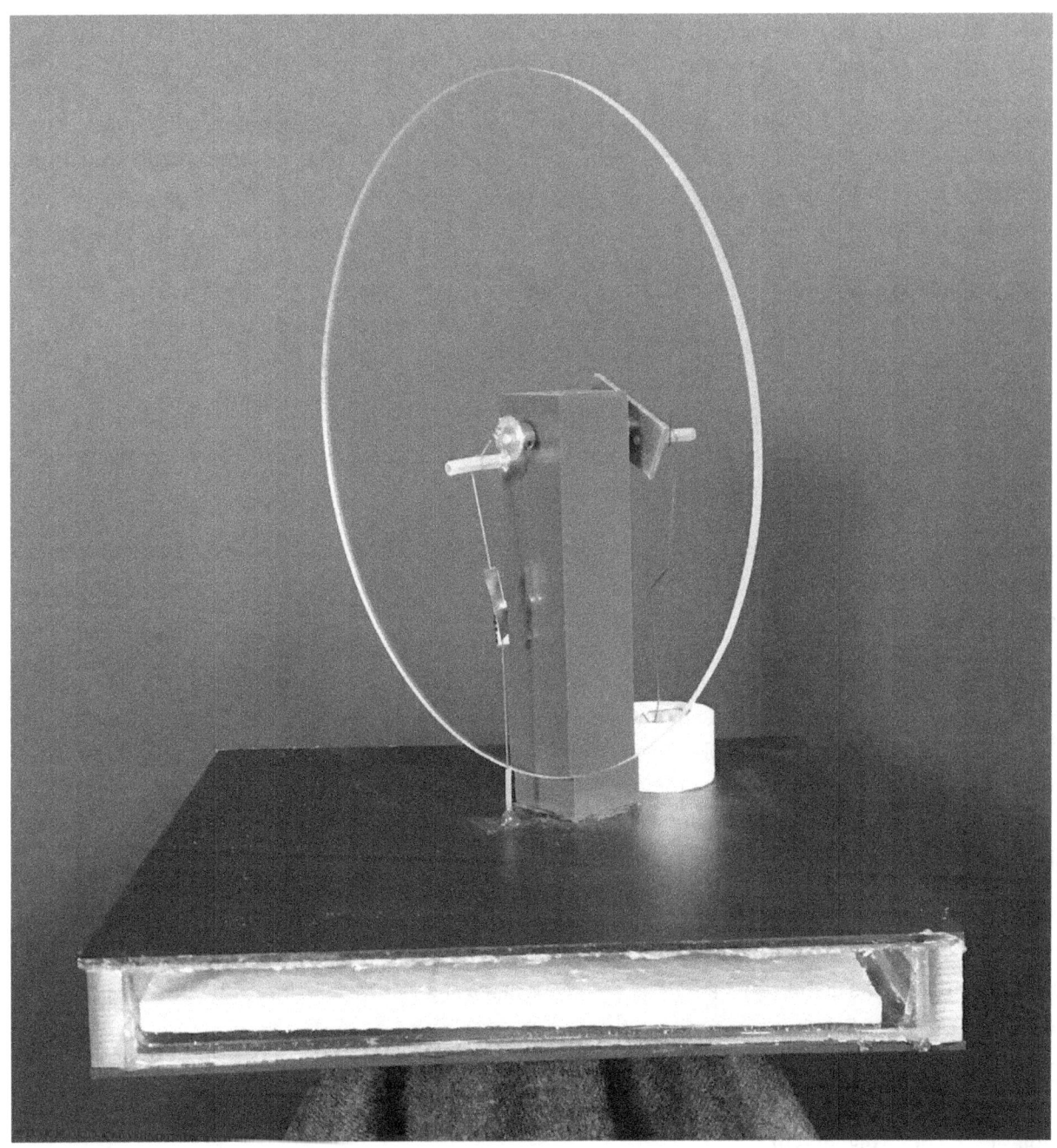

The finished 6″ (15.2 cm) square engine.

Trouble Shooting Tips

If the engine is not running well, it may be because of a problem in one of these four areas:

Temperature Differential: As mentioned previously, the engine should run with a temperature differential of 20° F (11° C). Setting the engine on a cup of near-boiling water in a 70° F (21° C) room should provide a temperature differential of about 100° F (56° C). If the engine is not running under these conditions, there are one or more other problems that will need to be fixed, such as a small pressure leak or friction.

It may be possible to overcome a small pressure leak or friction by increasing the temperature differential. This will increase the power output of the engine and possibly overcome a small amount of friction or a small pressure leak.

Increase the temperature differential by adding ice to the top of the motor while the bottom is being warmed by the heat source. Do not attempt to add more heat, as this can damage the engine. The Styrofoam displacer material may melt if the heat source is too hot.

Pressure Leaks: It does not take much of a leak to prevent the engine from running well. There are a couple of ways to test for a pressure leak. The first method is to observe the behavior of the diaphragm when the engine is at running temperature.

Disconnect the drive diaphragm pushrod from the crank pin. Place the engine on a cup of hot water and wait a few moments for the bottom side to heat up. Now, rotate the flywheel so that the displacer rises and falls inside the pressure chamber and observe the motion of the drive diaphragm. The diaphragm should move up and down in response to the heating and cooling of the air inside the pressure chamber. If this motion is not present, or if it is very limited, there may be a pressure leak. It may also be possible that the tension of the diaphragm is too tight or too loose.

The other method for leak testing also involves removing the diaphragm pushrod from the crank pin. Once it is disconnected, pull upward on the pushrod for 5 to 10 seconds to inflate the diaphragm. Release the pushrod and observe the diaphragm. If it immediately deflates and returns to a low or neutral position, there may be a pressure leak.

Pressure leaks can happen at a number of places:

- Holes in the drive diaphragm
- Leaking around the edge of the drive diaphragm
- Leaking glue joints in the pressure chamber
- Leaking through excessive clearance around the displacer pushrod.

It may take a bit of detective work to find the leak. It may be possible to patch a small leak with a small drop of glue or silicone sealant, or it may be necessary to replace the defective part.

Friction: Small amounts of friction can have a huge impact on an LTD Stirling engine's ability to run. Friction occurs at every point where two moving parts touch, and at every point where a moving part contacts the atmosphere. In the micro-horsepower world of LTD Stirling engines, a tiny bit of friction can stop the engine from performing.

Check the rotation of the axle by removing the connections to the displacer and the drive diaphragm and spinning the flywheel. The flywheel should coast to a stop after about 30 seconds after receiving a good spin by hand. If the flywheel does not spin freely there is a problem with friction somewhere in the axle assembly. Locate the cause of the friction and repair the problem.

The flywheel rotation should be smooth and silent during the spin test. Vibration and noise are both indications of friction.

Crankshaft Timing: There must be a 90^0 phase difference between the two crank pins. This means that when one pin is in the 12:00 o'clock position, the other one is at either 9:00 o'clock or at 3:00 o'clock. The engine will run with a phase difference in either direction. The only difference will be the direction the engine rotates while running.

When the bottom of the engine is heated, the engine will run with the motion of the displacer moving ahead of the drive diaphragm. This means that when the displacer is at the top of the pressure chamber, the drive diaphragm is halfway up and moving in an upwards direction. As the flywheel rotates and the displacer comes to the lowest point in its travel, the drive diaphragm is halfway down and moving downward. It is important to know which way the engine will run so that the initial push to get it started is in the same direction.

Engine #6: Small Round Engine 4" (10.2 cm)

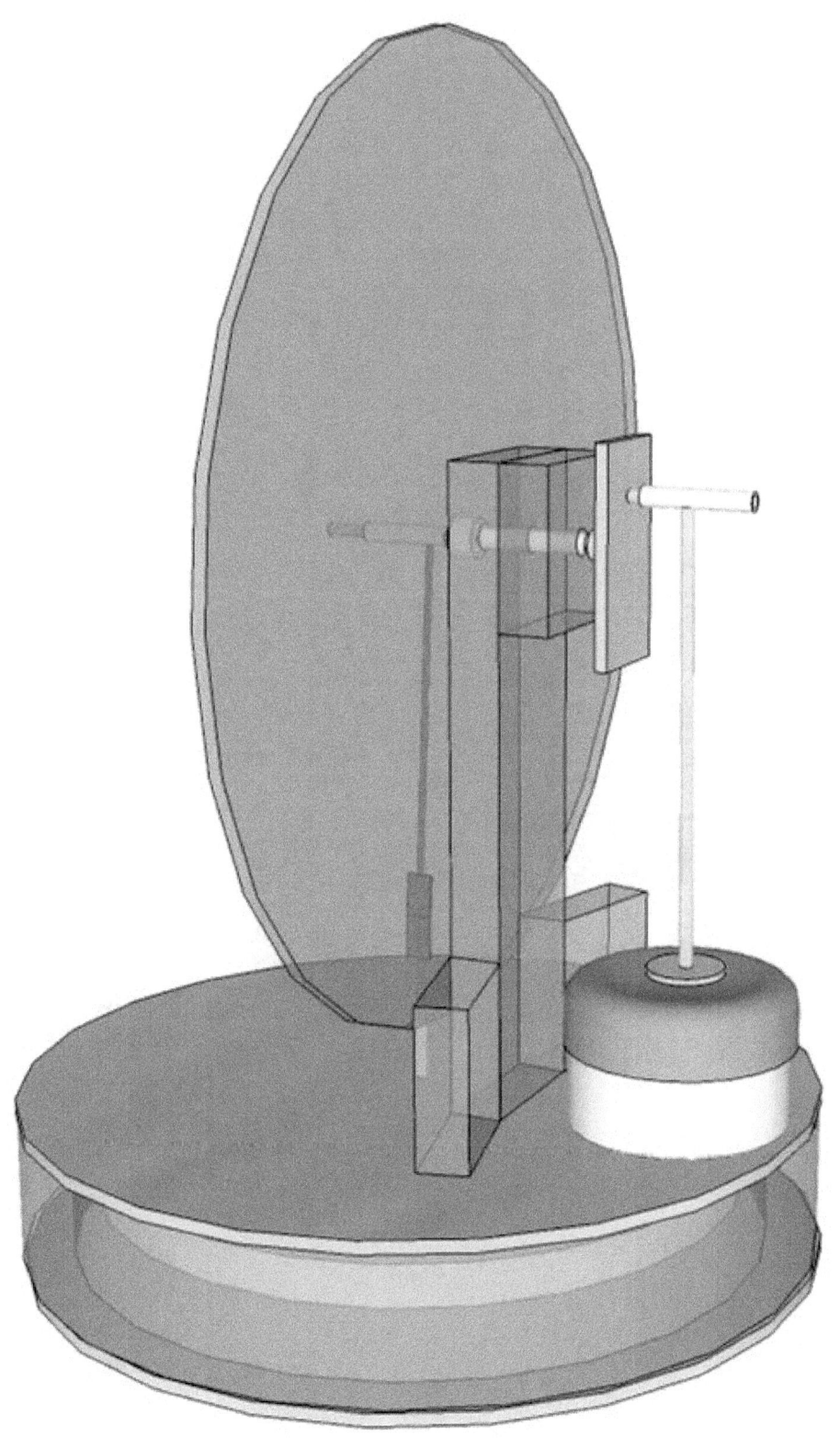

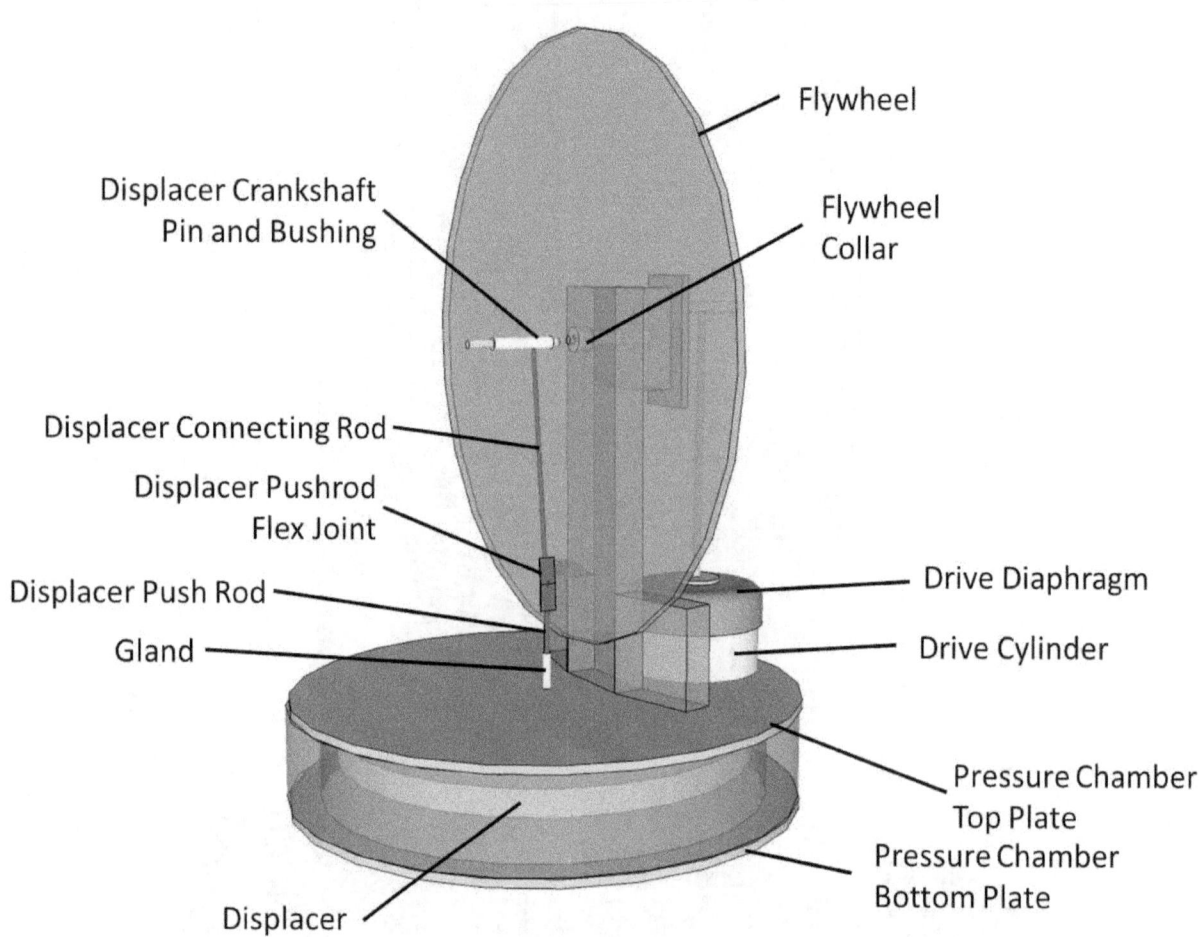

Flywheel

Displacer Crankshaft
Pin and Bushing

Flywheel
Collar

Displacer Connecting Rod

Displacer Pushrod
Flex Joint

Displacer Push Rod

Gland

Drive Diaphragm

Drive Cylinder

Pressure Chamber
Top Plate
Pressure Chamber
Bottom Plate

Displacer

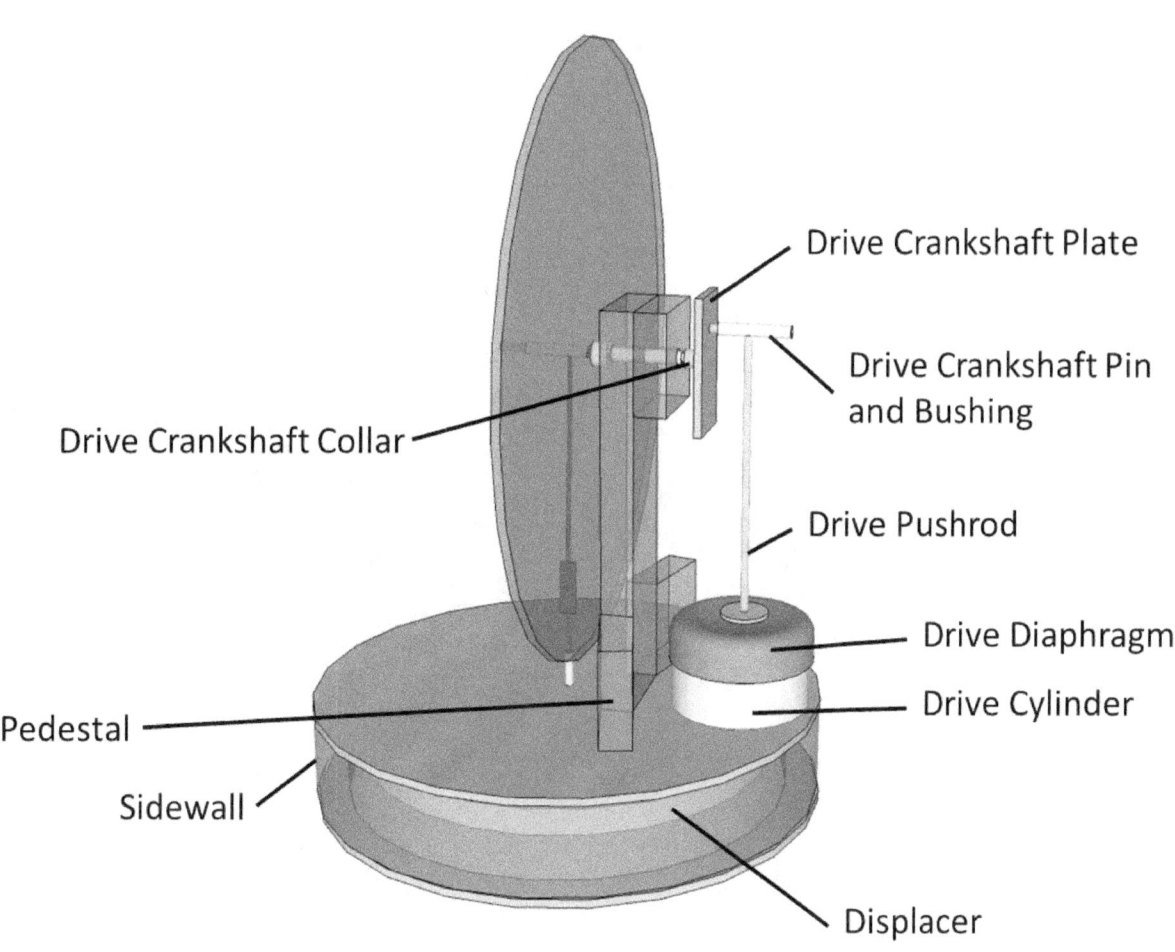

Drive Crankshaft Plate

Drive Crankshaft Pin and Bushing

Drive Crankshaft Collar

Drive Pushrod

Drive Diaphragm

Drive Cylinder

Pedestal

Sidewall

Displacer

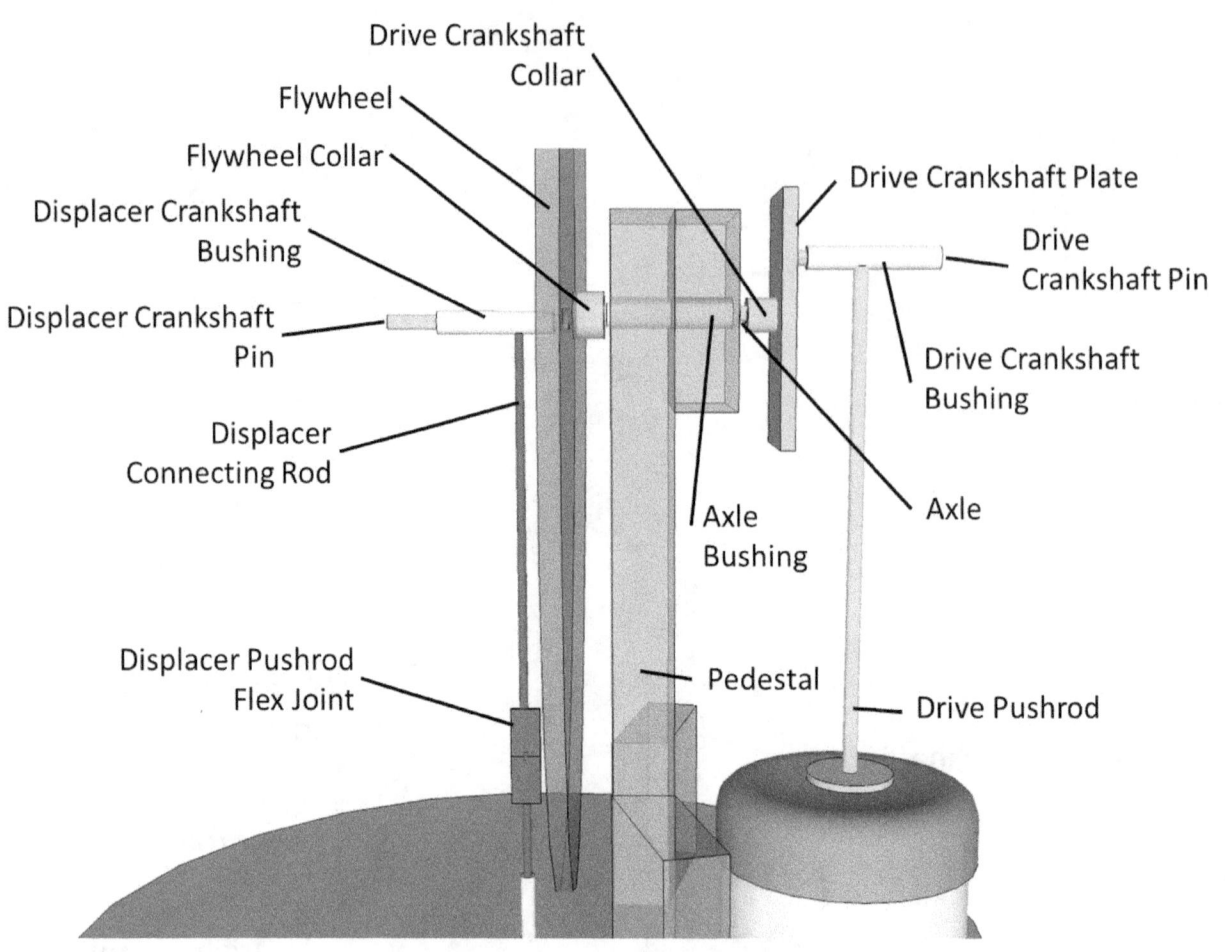

Drive Crankshaft Collar

Flywheel

Flywheel Collar

Displacer Crankshaft Bushing

Displacer Crankshaft Pin

Displacer Connecting Rod

Displacer Pushrod Flex Joint

Drive Crankshaft Plate

Drive Crankshaft Pin

Drive Crankshaft Bushing

Axle

Axle Bushing

Pedestal

Drive Pushrod

Parts and Materials

Part #	Part Name	Description
01	Flywheel	Clear Acrylic Sheet, round, 0.06" x 4 1/2" (11.43 cm) diameter*
02	Flywheel Collar	Shaft Collar, 1/16" (1.59 mm)
03	Pressure Chamber Bottom Plate	Aluminum Sheet, round, 0.062" x 4" (10.16 cm) diameter
04	Pressure Chamber Top Plate	Aluminum Sheet, round, 0.062" x 4" (10.16 cm) diameter
05	Displacer	Styrofoam, round, 0.20" (5 mm) x 3 3/8" (8.57 cm) diameter
06	Displacer Crankshaft Pin	Music Wire, 0.0625 x 1/2" (12.7 mm)
07	Displacer Crankshaft Bushing	AWG-14 (0.066") Heavy Wall Extruded Teflon Tube, 3/8" (9.53 mm)
08	Pressure Chamber Sidewall	Formed Clear Acrylic, 1/4" (6.35 mm)** x 11/16" (17.46 mm) x approx. 13" (33.02 cm)
09	Pedestal	Clear Acrylic, 1/4" (6.35 mm) x 5/8" (15.88 mm) x 3" (7.62 cm)
10	Pedestal Supports (3 pieces)	Clear Acrylic, 1/4" (6.35 mm) x 5/8" (15.88 mm) x 5/8" (15.88 mm), 3 pieces
11	Displacer Connecting Rod	Music Wire, 0.015" x 3 1/2" (8.89 cm)
12	Displacer Pushrod	Music Wire 0.015" x 6" (15.24 cm)
13	Displacer Pushrod Flex Joint	Duct Tape, 1/8" (3.18 mm) x 1/2" (12.7 mm), 2 pieces
14	Drive Crankshaft Pin	Music Wire, 0.0625 x 1/2" (12.7 mm)
15	Drive Crankshaft Bushing	AWG-14 (0.066") Heavy Wall Extruded Teflon Tube, 3/8" (9.53 mm)
16	Drive Crankshaft Collar	Shaft Collar, 1/16" (1.59 mm)
17	Drive Crankshaft Plate	Aluminum Sheet, 0.062" 1/4" (6.35 mm) x 3/8" (9.53 mm)
18	Gland	Teflon Tube, #24 (0.022" ID) x 7/16" (11.11 mm)
19	Drive Cylinder	PVC Pipe, 1" (25.4 mm) ID x 5/8" (15.88 mm)
20	Drive Diaphragm	Latex glove fingertip
21	Drive Pushrod	Music Wire, 0.015" x 3 1/2" (8.89 cm)
22	Axle	Music Wire, 0.0625"x 1" (25.4 mm)
23	Axle Bushings (2 pieces)	AWG-14 (0.066") Heavy Wall Extruded Teflon Tube, 3/16" (4.76 mm)

*Flywheel thickness can vary slightly from the prescribed thickness of 0.06", depending upon the material available to the builder. Acrylic sheet commonly sold in the US market for window glazing is slightly thicker (0.093") and is used for the flywheels of some of the engines in this book. Thin polystyrene sheet material used in covering framed pictures is only 0.05" thick and will suffice, but can be difficult to work with.

** It is also possible to make the sidewalls from slightly thinner 3/16" (4.76 mm) acrylic sheet.

Part #8 is a round sidewall that is formed by heating a straight piece of acrylic sheet and bending it around a form.

Drawings and Dimensions

Note: Images are not drawn to scale.

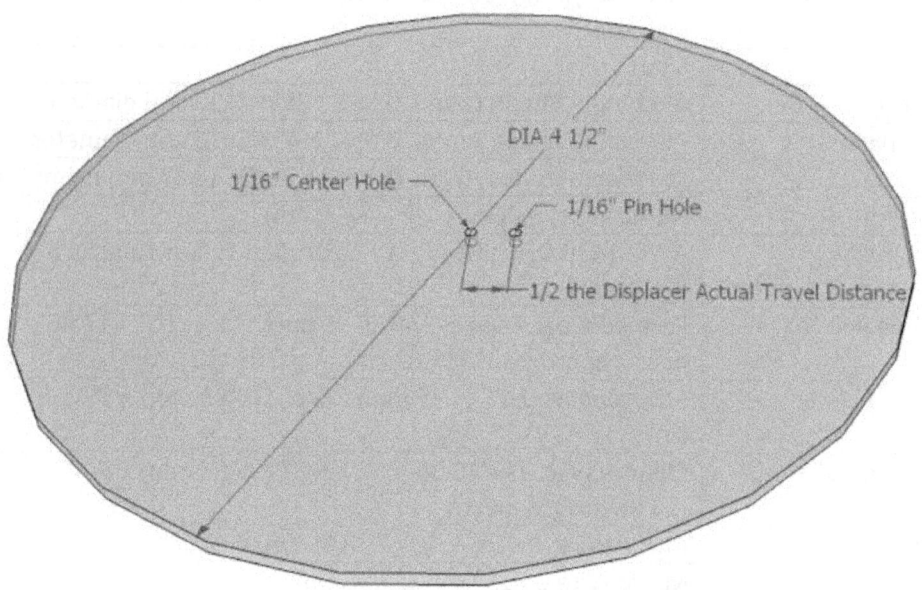

Part # 01 Flywheel - Clear Acrylic Sheet, round, 0.06" x 4 1/2" (11.43 cm) diameter

Part # 02 Flywheel Collar - 1/16" (1.59 mm) Shaft Collar
Part # 16 Drive Crankshaft Collar - 1/16" (1.59 mm) Shaft Collar

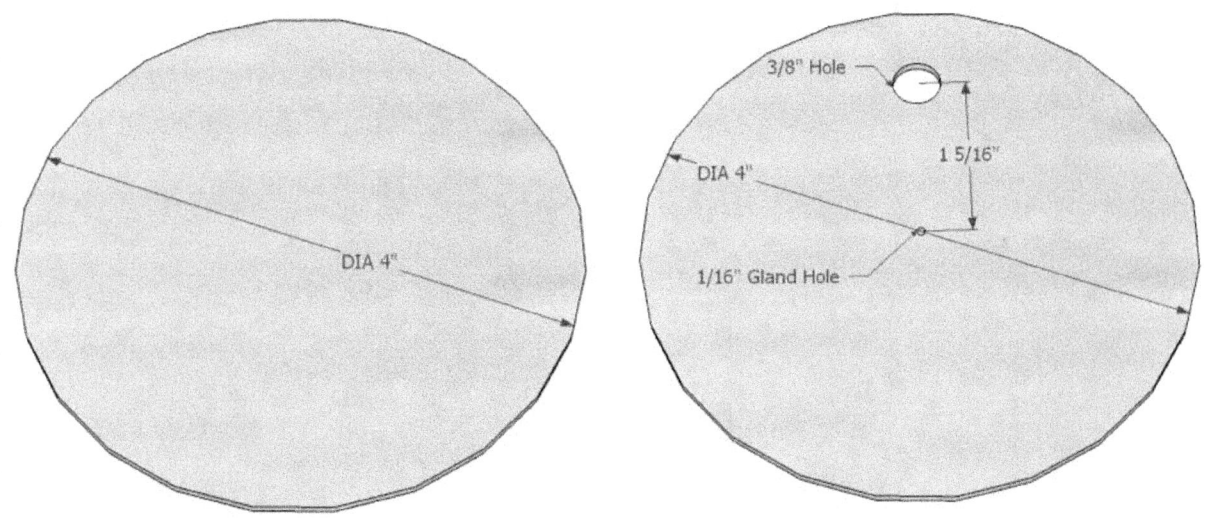

Part # 03 Pressure Chamber Bottom Plate   **Part # 04 Pressure Chamber Top Plate**

Aluminum Sheet, 0.062" x 4" (10.16 cm) x 4" (10.16 cm) (2 pieces)

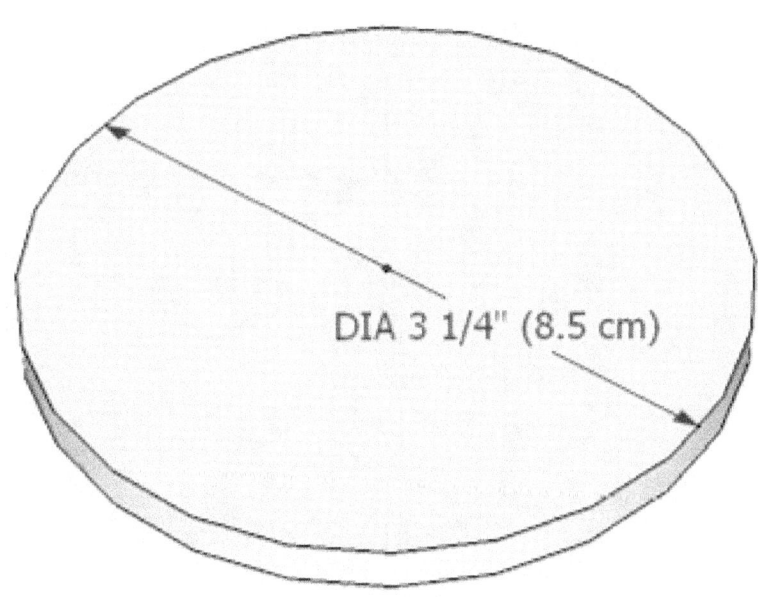

Part # 05 Displacer - Styrofoam, 0.20" x 3 1/4" (8.5 cm)

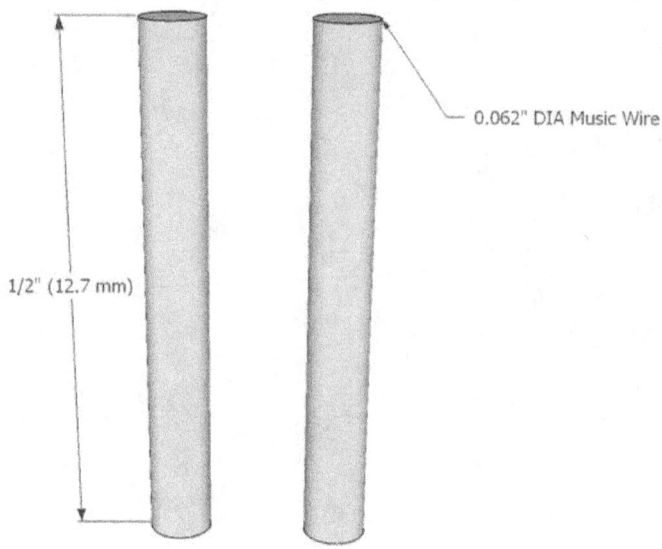

0.062" DIA Music Wire

1/2" (12.7 mm)

Part # 06 Displacer Crankshaft Pin - Music Wire, 0.0625 x 1/2" (12.7 mm)
Part # 14 Drive Crankshaft Pin - Music Wire, 0.0625 x 1/2" (12.7 mm)

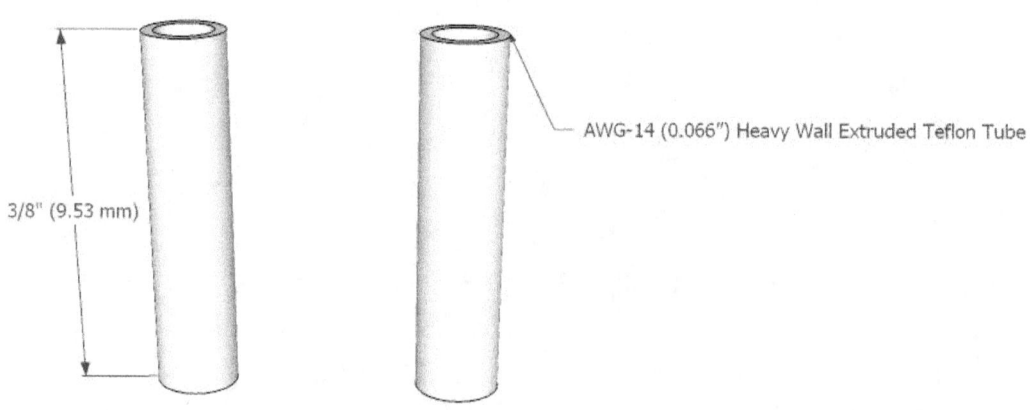

AWG-14 (0.066") Heavy Wall Extruded Teflon Tube

3/8" (9.53 mm)

Part # 07 Displacer Crankshaft Bushing - AWG-14 (0.066") Heavy Wall Extruded Teflon Tube, 3/8" (9.53 mm)
Part # 15 Drive Crankshaft Bushing - AWG-14 (0.066") Heavy Wall Extruded Teflon Tube, 3/8" (9.53 mm)

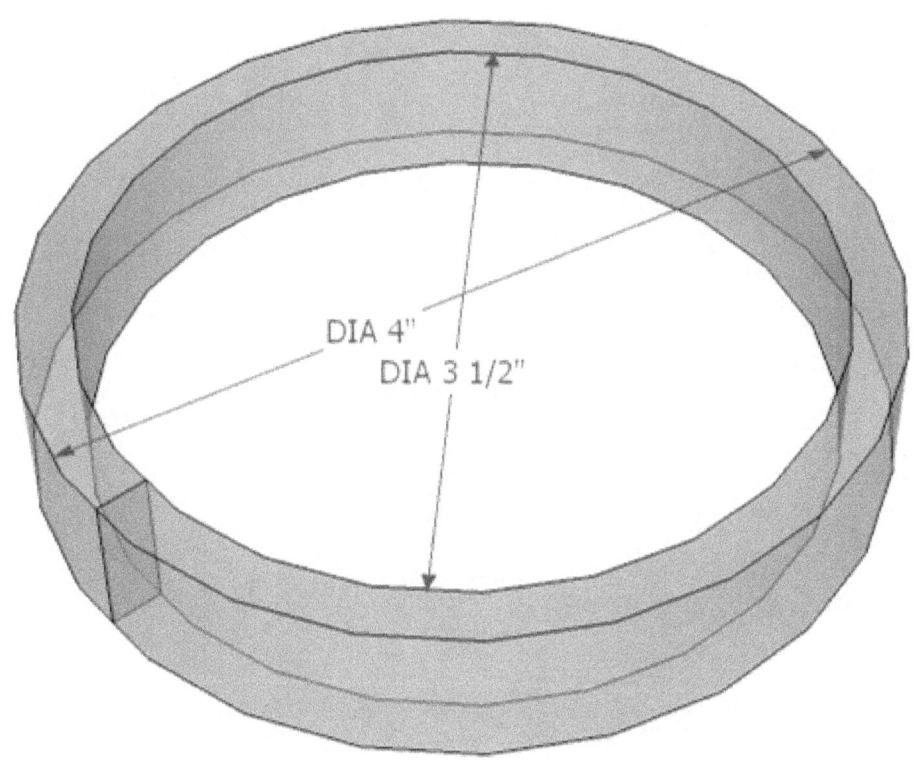

DIA 4"
DIA 3 1/2"

Part # 08 Pressure Chamber Sidewall - Formed Clear Acrylic, 1/4" (6.35 mm)** x 11/16" (17.46 mm) x approx. 13" (33.02 cm). Finished dimensions: OD 4" (10.16 cm), ID 3 1/2" (8.89 cm)

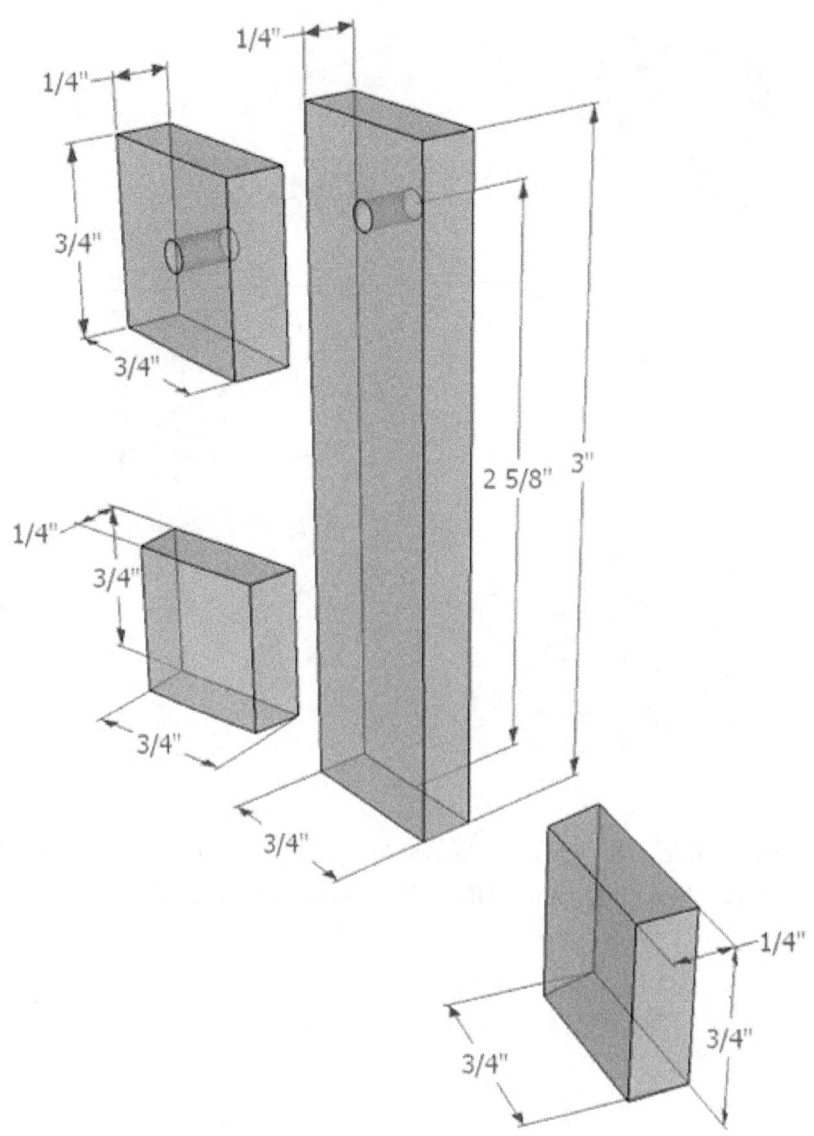

Part # 09 Pedestal - Clear Acrylic, 1/4" (6.35 mm) x 5/8" (15.88 mm) x 3" (7.62 cm)

Part # 10 Pedestal Supports- Clear Acrylic, 1/4" (6.35 mm) x 5/8" (15.88 mm) x 5/8" (15.88 mm) (3 pieces)

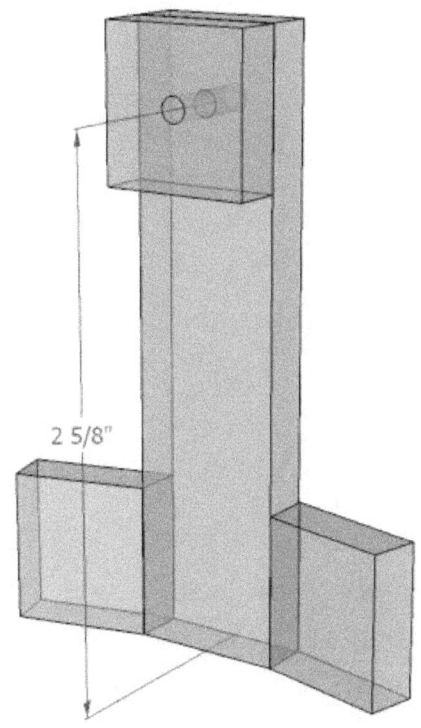

2 5/8"

Part # 09 Pedestal – and **Part # 10 Pedestal Supports**- Assembled

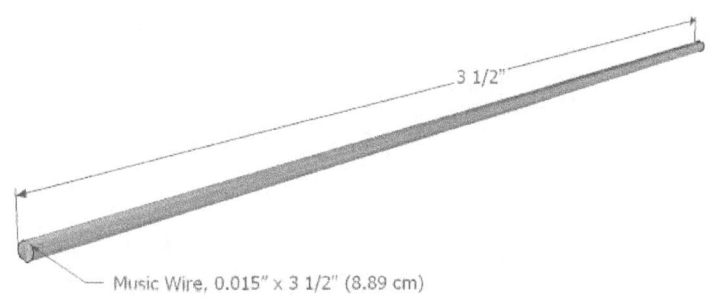

3 1/2"

Music Wire, 0.015" x 3 1/2" (8.89 cm)

Part # 11 Displacer Connecting Rod - Music Wire, 0.015" x 3 1/2" (8.89 cm)
Part # 21 Drive Pushrod - Music Wire, 0.015" x 3 1/2" (8.89 cm)
(2 Pieces)

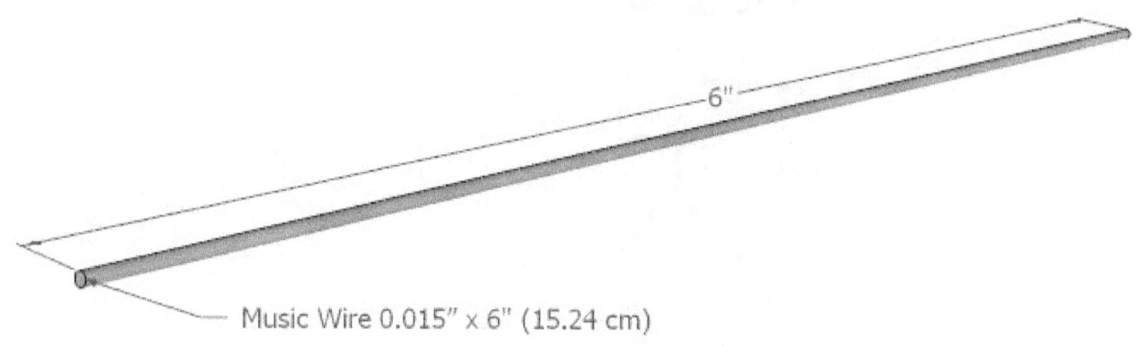

Music Wire 0.015" x 6" (15.24 cm)

Part # 12 Displacer Pushrod - Music Wire 0.015" x 6" (15.24 cm)

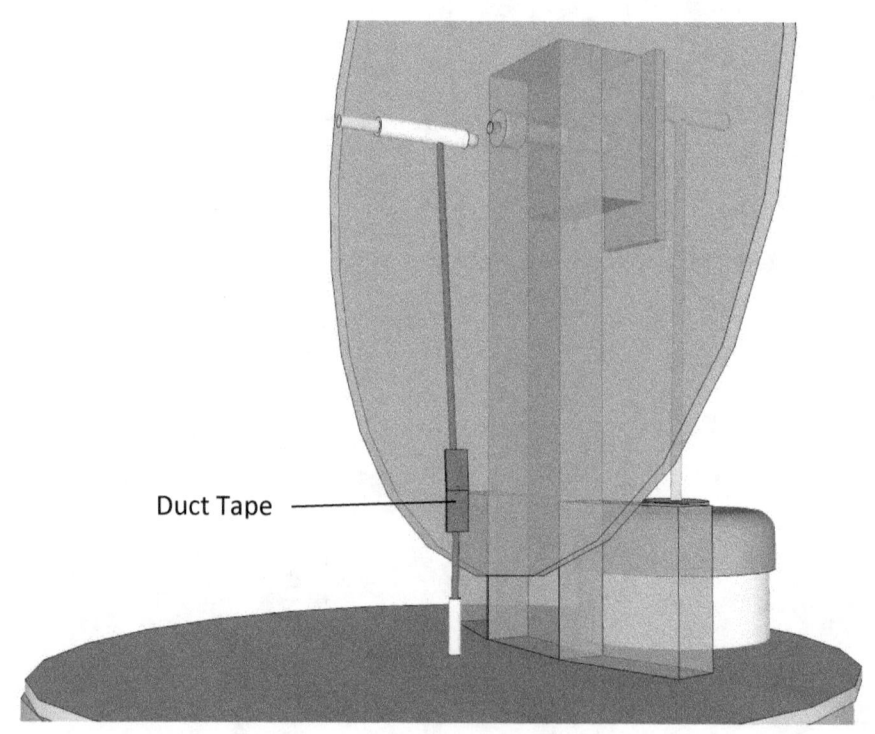

Duct Tape

Part # 13 Displacer Pushrod Flex Joint - Duct Tape, 1/8" (3.18 mm) x 1/2" (12.7 mm) (2 pieces)

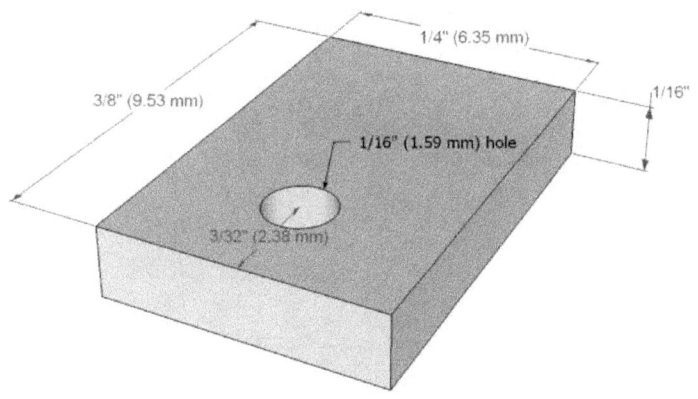

Part # 17 Drive Crankshaft Plate - Aluminum Sheet, 0.062" (1/16", 1.59 mm), 1/4" (6.35 mm) x 3/8" (9.53 mm)

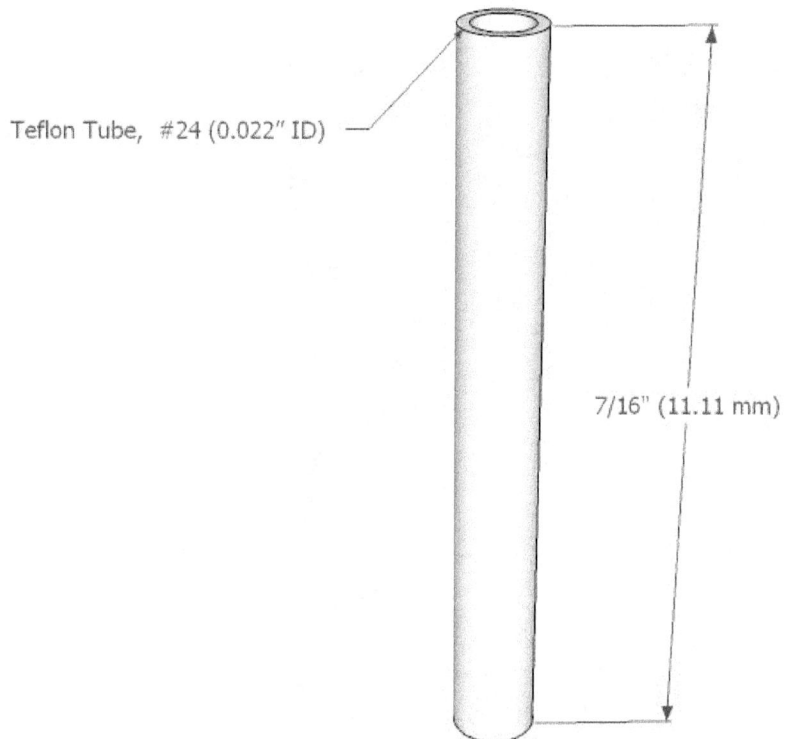

Part # 18 Gland - Teflon Tube, #24 (0.022" ID) x 7/16" (11.11 mm)

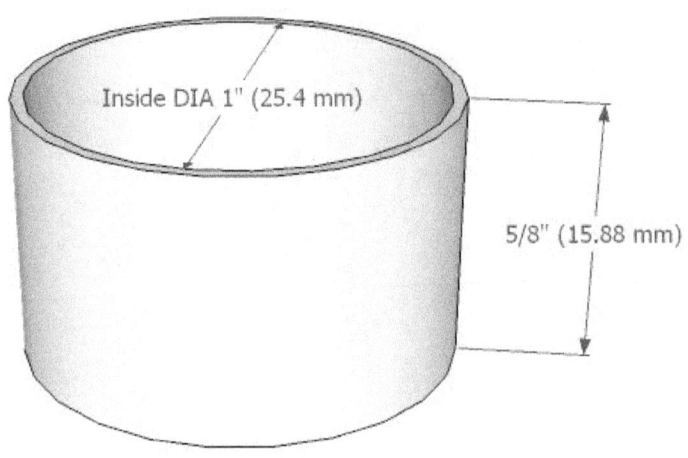

Part # 19 Drive Cylinder - PVC Pipe, 1" (25.4 mm) ID x 5/8" (15.88 mm)

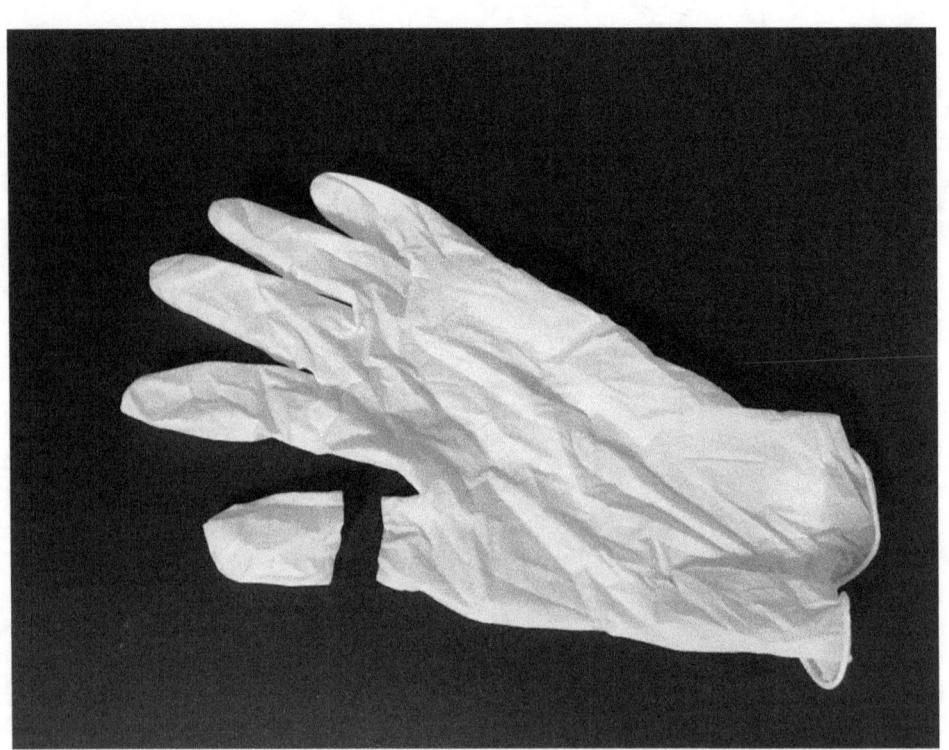

Part # 20 Drive Diaphragm - Latex glove fingertip

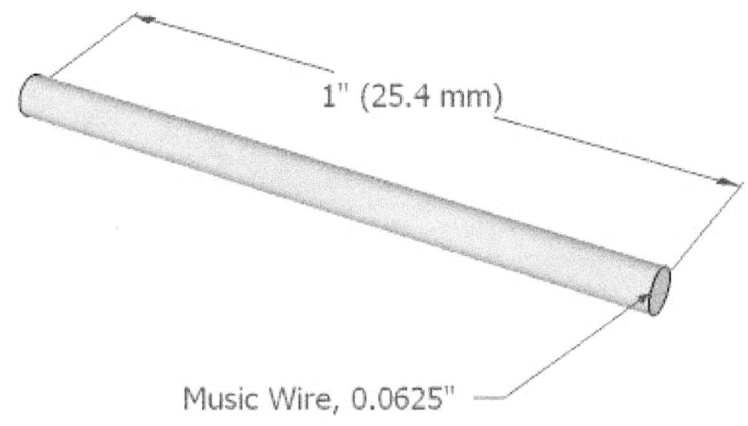

Part # 22 Axle - Music Wire, 0.0625"x 1" (25.4 mm)

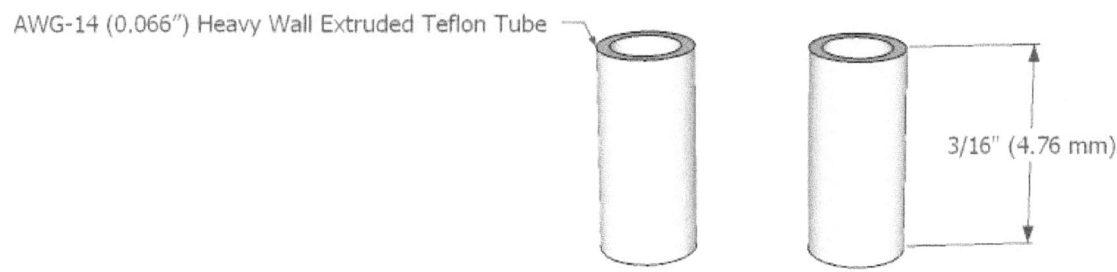

Part # 23 Axle Bushings (2 pieces) - AWG-14 (0.066") Heavy Wall Extruded Teflon Tube, 3/16" (4.76 mm)

Assembly Instructions for Small Round Engine

Make the Pressure Chamber Top and Bottom Plates

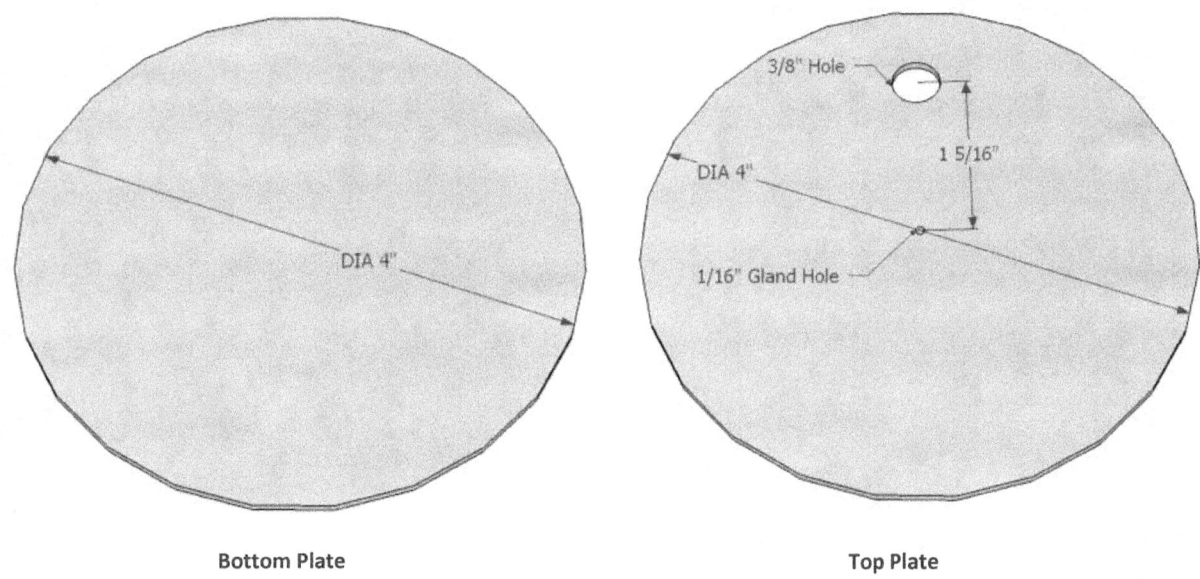

Bottom Plate   Top Plate

Cut two aluminum plates to form the top and bottom plates of the pressure chamber. Plate thickness is 0.0625" (1.59 mm). The plates are round, measuring 4" (10.16 cm) in diameter

Wash the metal parts with soap and water, rinse well and allow them to dry.

Paint all surfaces of the top and bottom plates (if desired). Use spray enamel that is suitable for painting metal.

Mark the center point of the top plate when laying out the pattern on the material. Use a center punch to mark the spot. Drill a hole that is the same diameter (or slightly larger) than the outside diameter of the Teflon tube that will be used for the displacer gland. The size will vary depending upon the material chosen.

Mark the position for the drive cylinder hole. The hole is located 1 5/16" (33.34 mm) from the center of the top plate. The size of the hole is 3/8" (10 mm). Mark the position for the hole with a center punch and then carefully drill the hole.

The top plates for these engines were cut with a band saw.

Make the Pressure Chamber Side Wall

The pressure chamber side wall is made by heating and bending a piece of clear acrylic. This process is sometimes called thermoforming. This process will require these three things:

- A piece of clear acrylic sheet measuring 1/4" (6.35 mm) x 11/16" (17.46 mm) x approximately 13" (33.02 cm) long.
- A heat gun to warm the acrylic.
- A jig or form to shape the acrylic and hold it in the proper shape until it cools.

Cut the Material to Length

The sidewall can be made from a single piece of acrylic, or it can be made in sections and pieced together. The examples shown here were made in sections and pieced together because the material available was not long enough to form the side from a single piece.

The length of the sidewall will be the same as the circumference of the top and bottom plates. The circumference of the plates can be calculated by multiplying the diameter of the circle by 3.1416.

4" x 3.1416 = 12.57", or 12 9/16"

10.16 cm x 3.1416 = 31.93 cm

140

The acrylic sheet material should be about 1/4" (6.35 mm) thick. Thinner material may be substituted. The thickness of the material used to make the sidewall will determine the size of the form that needs to be made for shaping the material during thermoforming.

Build a Form

The form is a round disk that is the same size as the inside diameter of the finished sidewall. This dimension is calculated by subtracting the thickness of the sidewall material (times 2) from the diameter of the pressure chamber. So for 1/4" (6.35 mm) material the inside diameter of the sidewall will be 3 1/2" (8.89 cm).

This picture shows a form built for bending the sidewall to make the 4" (10.16 cm) engine. The holes make it possible to use spring clamps to hold the material against the form.

Shape the Acrylic Sidewall with Heat

A heat gun is used to warm the acrylic sheet until it becomes soft and pliable. Use extreme caution and work in a well ventilated area. Acrylic will generate poisonous fumes if overheated, and it is flammable. Use just enough heat to soften and bend the acrylic, and no more. The heat gun is hot enough to cause severe burns or fire, so use caution with this tool.

Use a clamp to hold one end of the acrylic strip against the form. Apply heat to the first several inches of material. Apply occasional light pressure against the material to test it. It will eventually become soft with a consistency similar to clay.

Continue to apply heat ahead of the area to be bent as the acrylic is turned around the shape of the form.

This picture shows a strip of acrylic sheet as it is clamped in place in preparation for bending. The protective covering will have to be removed from the acrylic sheet before heat is applied.

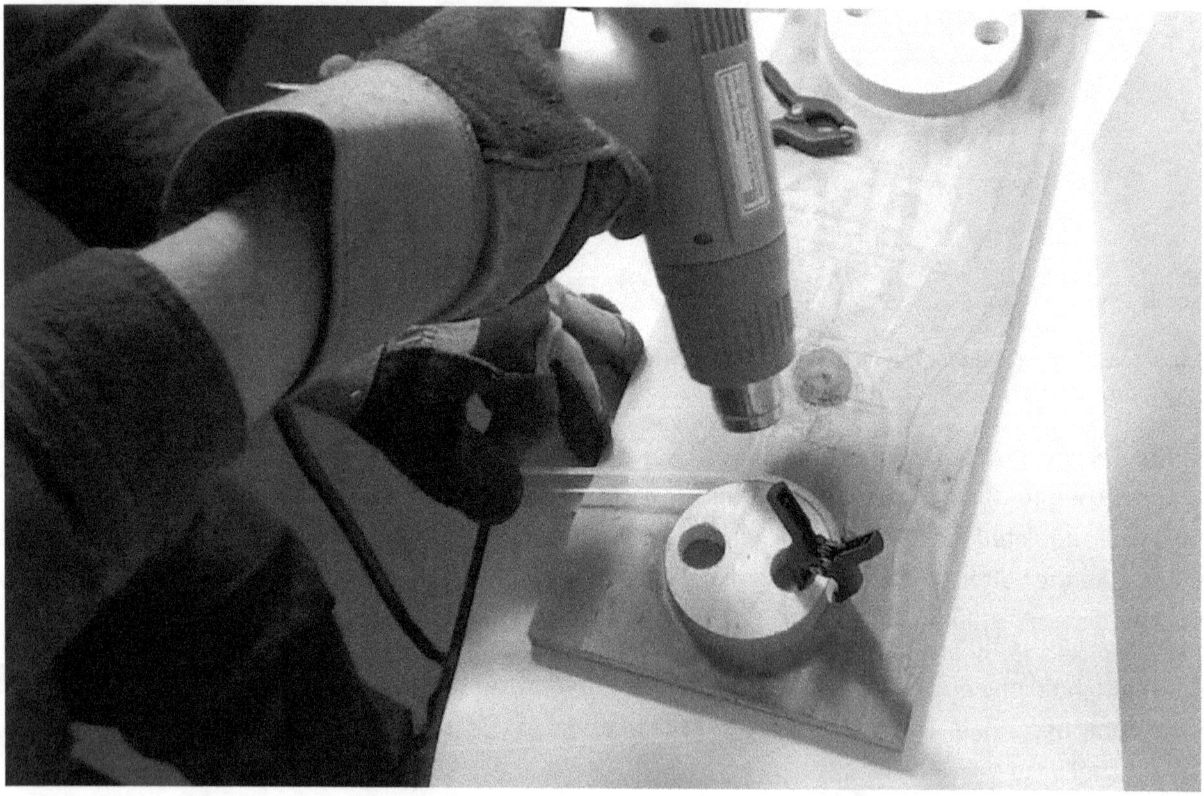

Apply heat to soften the material and work around the form.

Continue heating the material as it is bent to the shape of the form.

When finished, allow the material to remain against the form until the acrylic has cooled and become rigid again. Since this piece was not long enough to complete the circle, it will be joined with another piece to make a complete sidewall.

Attach the Sidewall to the Pressure Chamber Bottom Plate

Carefully check the fit of all the parts before assembly and use enough adhesive to ensure a good seal. If multiple pieces were made to form the sidewall, they can be joined together during this process of attaching them to the bottom plate. Use clear five-minute epoxy to attach the acrylic sidewall to the pressure chamber bottom plate.

The sidewall must be attached so that the pressure chamber is air-tight. A small bead of glue oozing out of the joint indicates that the joint is being sealed well. Too much oozing glue, however, may interfere with the movement of the displacer inside the pressure chamber.

The top plate will _not_ be glued on at this time. The top plate may be used to help hold the sidewall in place as the glue hardens.

Note: The glue will leave a permanent mark anywhere that it comes in contact with the acrylic. Be very careful not to create too many fingerprints on the acrylic during the gluing process, as these may be permanent.

Note: Silicone adhesive is another option for gluing the sidewalls to the pressure chamber. Silicone adhesives tolerate heat better than most clear epoxy glues, but they do not hold as well. Silicone is especially useful if there is a need to disassemble the engine for any reason.

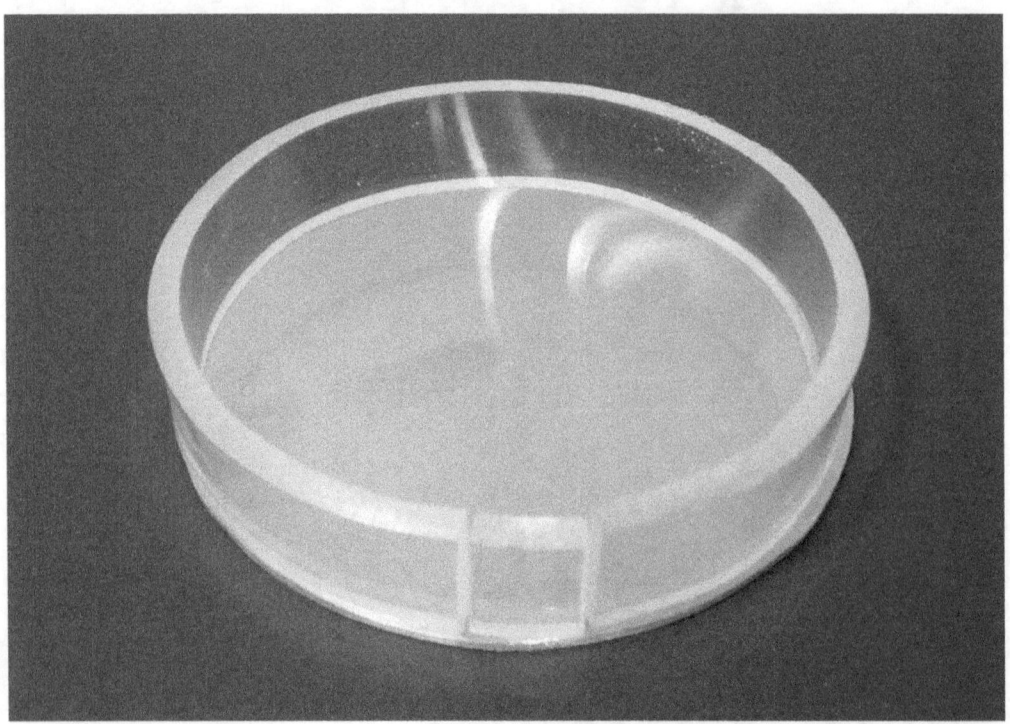

The pressure chamber bottom plate and sidewall pieces are ready for assembly. Dry fit the parts before gluing and make any adjustment necessary to create a good fit. This sidewall was made in two pieces because the stock was not long enough to make one continuous sidewall.

Cut the Foam Displacer Panel

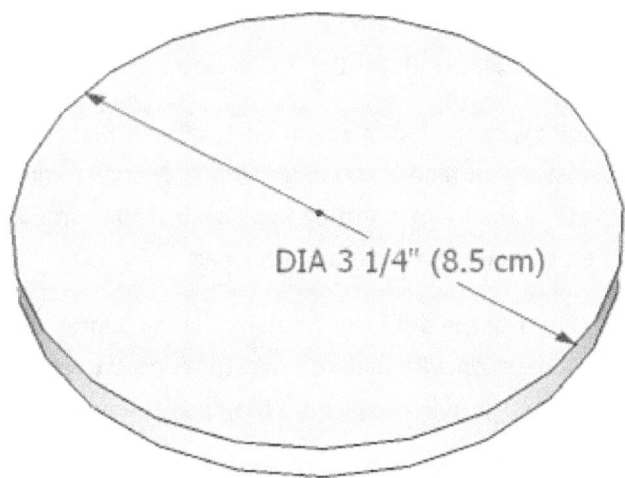

The displacer is made from Styrofoam that is cut to a thickness of 0.20" (13/64", 5.08 mm). For the best results, cut the foam with a hot wire foam cutter similar to the one described elsewhere in this book.

A hot wire foam cutter uses Nichrome wire that is heated by a small electrical current to cut the foam. The framework of the hot wire foam cutter holds tension on the wire to keep it straight, and a small electrical transformer provides the current to heat the wire. Foam cutters can be purchased from craft suppliers, or one can be built using simple parts. Plans for a homemade hot wire foam cutter are provided later in this book.

The dimensions of the displacer provide clearance between the displacer and the sidewall of the pressure chamber. That clearance needs to be 1/16" (1.59 mm) to 1/8" (3.18 mm) between the side of the displacer panel and the pressure chamber sidewall. Measure the internal dimensions of the pressure chamber and make any necessary adjustments to the size of the displacer before cutting.

Use a compass to create the circle on the foam sheet and mark the center of the circle. The center point will be used to locate the displacer pushrod.

Attach the Displacer Pushrod
Cut a piece of 0.015" music wire to a length of approximately 6" (15.24 cm). Small diameter music wire can be cut with ordinary wire cutters.

Make a 90° bend 1/2" from one end of the music wire.

A simple jig is used to hold the wire at the correct angle while attaching it to the foam displacer. Make the jig by drilling a small hole through a piece of flat wood. Drilling the hole in the jig will require a small diameter drill bit and a drill press. Choose a bit size that will provide a snug fit for the music wire.

If small drill bits under 1/16" (1.59 mm) are not available, use a 1/16" (1.59 mm) drill bit and insert a piece of Teflon tube into the hole to make a better fit for the music wire.

Pierce the foam at the center point with the long straight end of the displacer pushrod and press it in until the 90° bend is pressed up against the surface of the foam. Place the protruding music wire into the hole in the jig and pull it though the jig until the foam is setting flush against the surface of the jig. The jig should now be holding the pushrod perpendicular to the foam displacer.

Press down on the wire at the center of the displacer to make a small depression in the foam no more than 1/32" (0.79 mm) deep. Fill the depression with high temperature epoxy to attach the pushrod to the displacer. Leave the displacer and pushrod in the jig until the epoxy has cured.

The displacer pushrod is held at the correct angle by the wooden jig. A hole is drilled through the board to hold the shaft at the correct angle. High temperature epoxy is used to attach the pushrod to the displacer panel. This picture shows a square displacer. The process is the same for the round displacer.

Prepare the Drive Cylinder Tube

As noted earlier, the tubing used to make the drive cylinder is approximately 1" (25.4 mm) in diameter. Cut a piece of this tubing to a length of 5/8" (15.88 mm). Cut the pipe carefully to maintain a 90° angle on each end of the short drive cylinder.

Use fine sand paper to remove the glossy factory surface from the PVC pipe. This will also remove any labels and markings and improve the look of the finished part. Sanding the outer surface of the drive cylinder will help hold the latex drive diaphragm in place.

Place the drive cylinder over the hole near the edge of the pressure chamber top plate and glue it in place with epoxy. The drive cylinder is positioned flush to the edge of the pressure chamber top plate.

Apply the glue on the inside edge of the drive cylinder tube. Applying the glue on the inside edge will hide the glue joint from view when the engine is fully assembled.

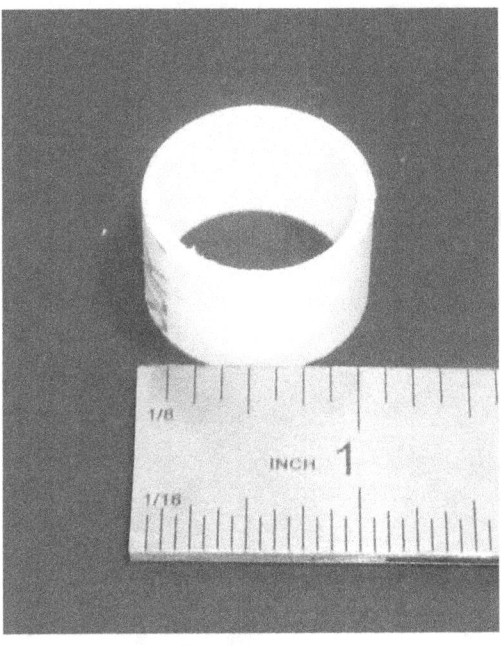

The drive cylinder is made from a short piece of 1"
(25.4 mm) diameter PVC pipe.

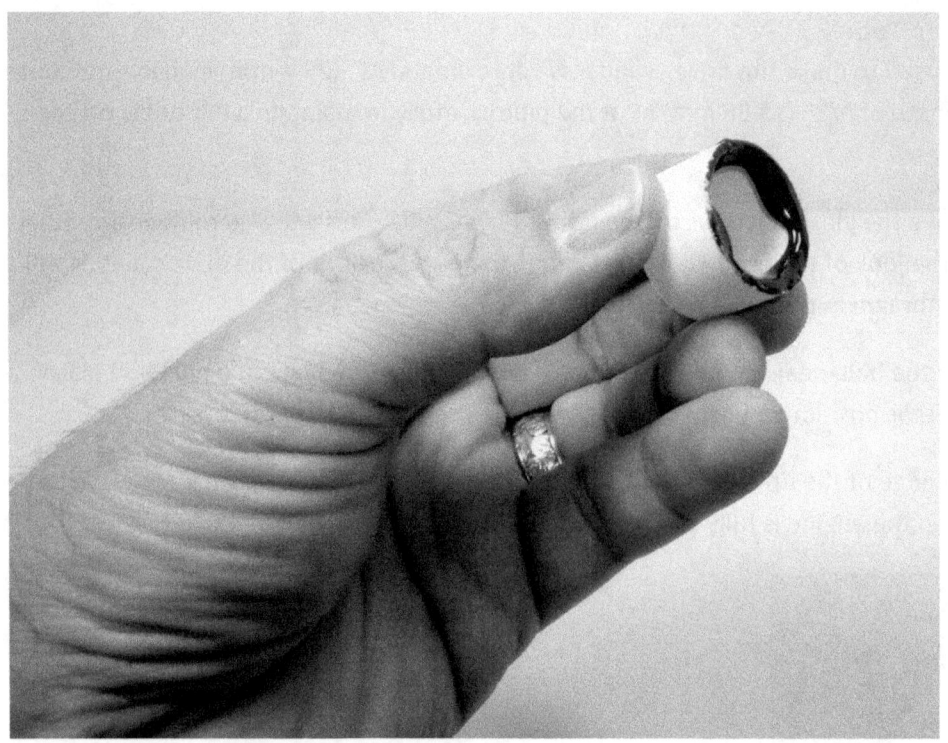

The drive cylinder is attached with epoxy. The glue is spread on the inside of the cylinder to create a nice looking airtight joint.

The drive cylinder is glued in place on the pressure chamber top. The glue is on the inside of the cylinder to provide an airtight seal that will be hidden from view after final assembly.

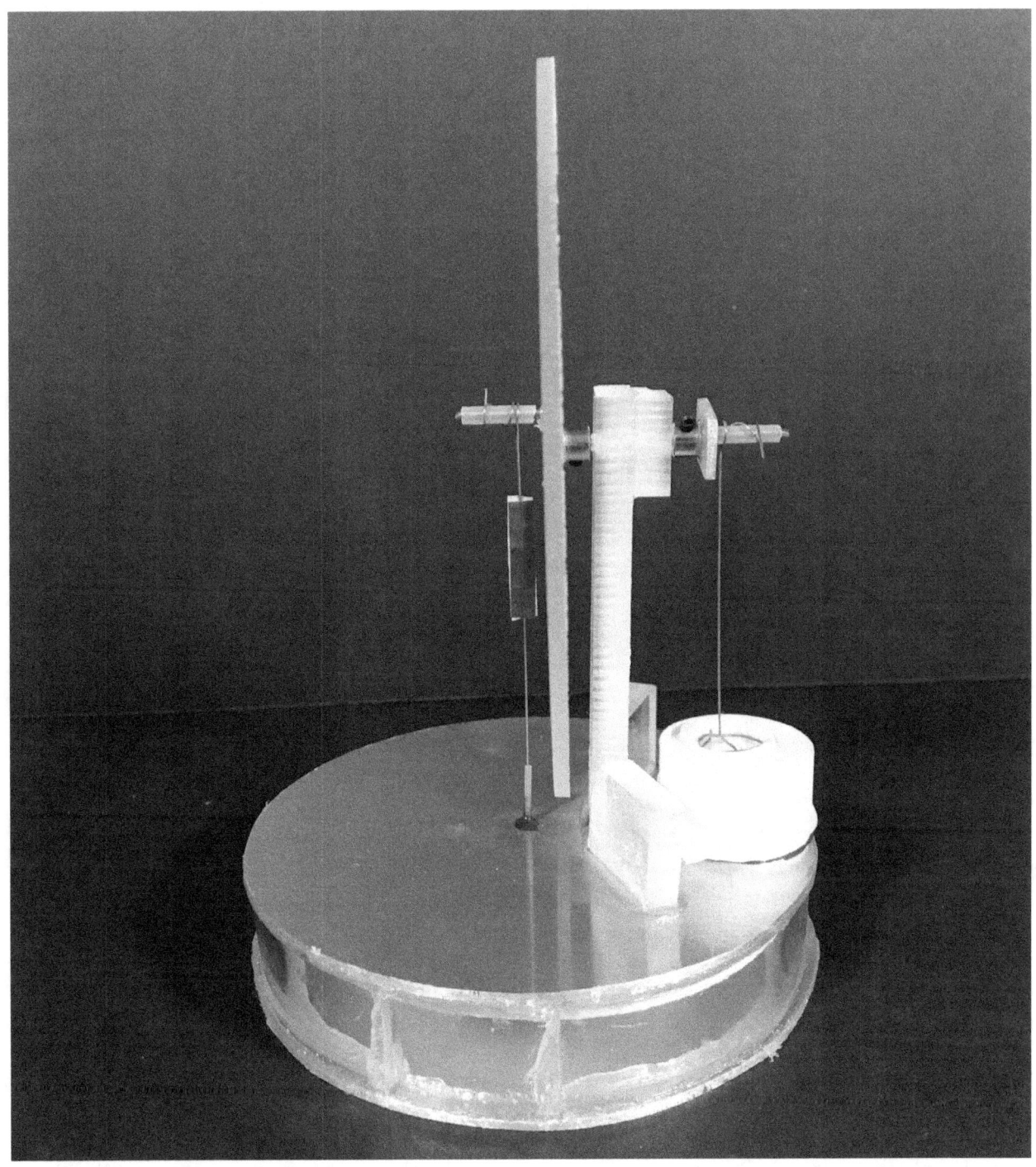

This picture shows a finished engine. Note the position of the drive cylinder at the edge of the top plate. The small size of this engine requires that the parts be small and close together.

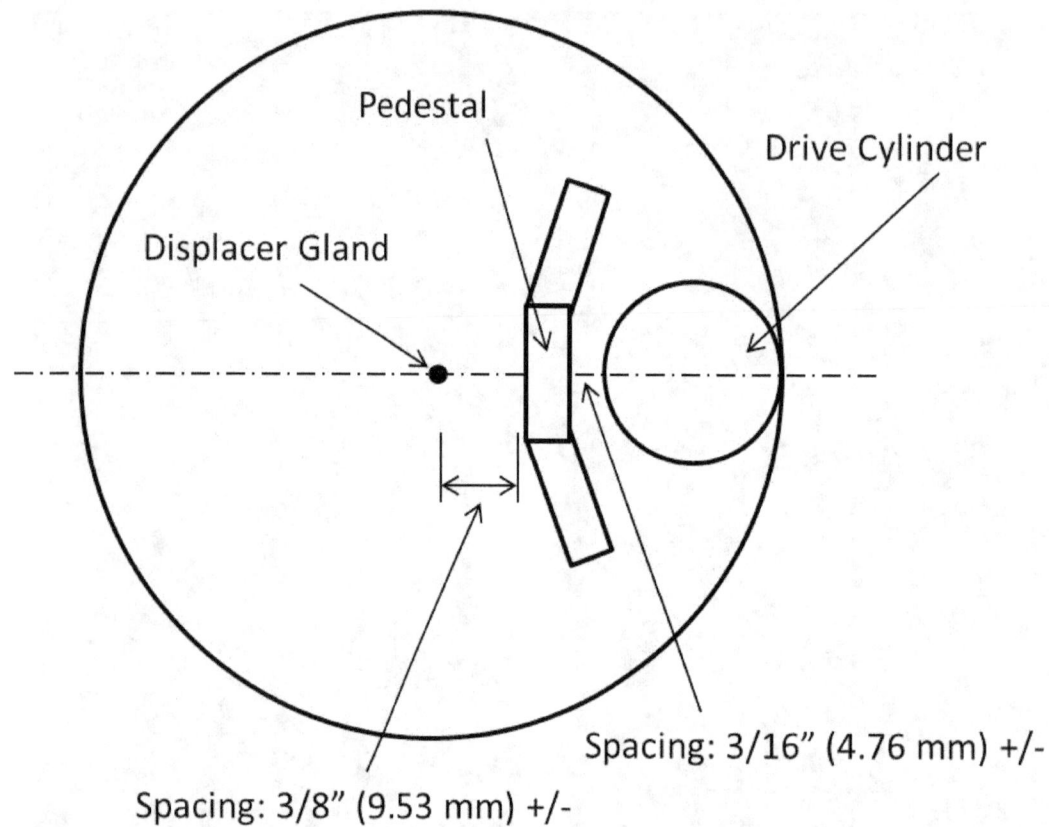

Pedestal

Drive Cylinder

Displacer Gland

Spacing: 3/16" (4.76 mm) +/-

Spacing: 3/8" (9.53 mm) +/-

This diagram illustrates the spacing of the parts on the top of the small round engine.

Attach the Displacer Gland

In the world of machinery, a "gland" is a sleeve within a stuffing box, fitted over a shaft in such a way as to prevent leakage of fluid while allowing a shaft or stem to move. Additionally, the gland of a Stirling engine allows the shaft to move with little or no friction. The gland is nearly air-tight. A tiny (very tiny!) pressure leak is desirable and helps the engine adjust for pressure changes as it warms up.

The gland is made of #24 (0.022" ID) Teflon tube. This provides a friction free fit for the 0.015" displacer pushrod.

Cut a piece of this Teflon tube to a length of 7/16" (11.11 mm).

Cutting Teflon tube requires a very sharp knife or razor blade, and a piece of 0.015" music wire. Insert the music wire into the Teflon tube before cutting. Place the tubing and wire on a hard flat surface. Press the blade of the knife against the tubing and roll the tubing on the flat surface until the knife has cut all the way around the tube. The wire in the middle of the tube will stop the knife from cutting all the way through the tube unless the tube is rolled. This will prevent the end of the tube from being crushed or deformed during the cutting process.

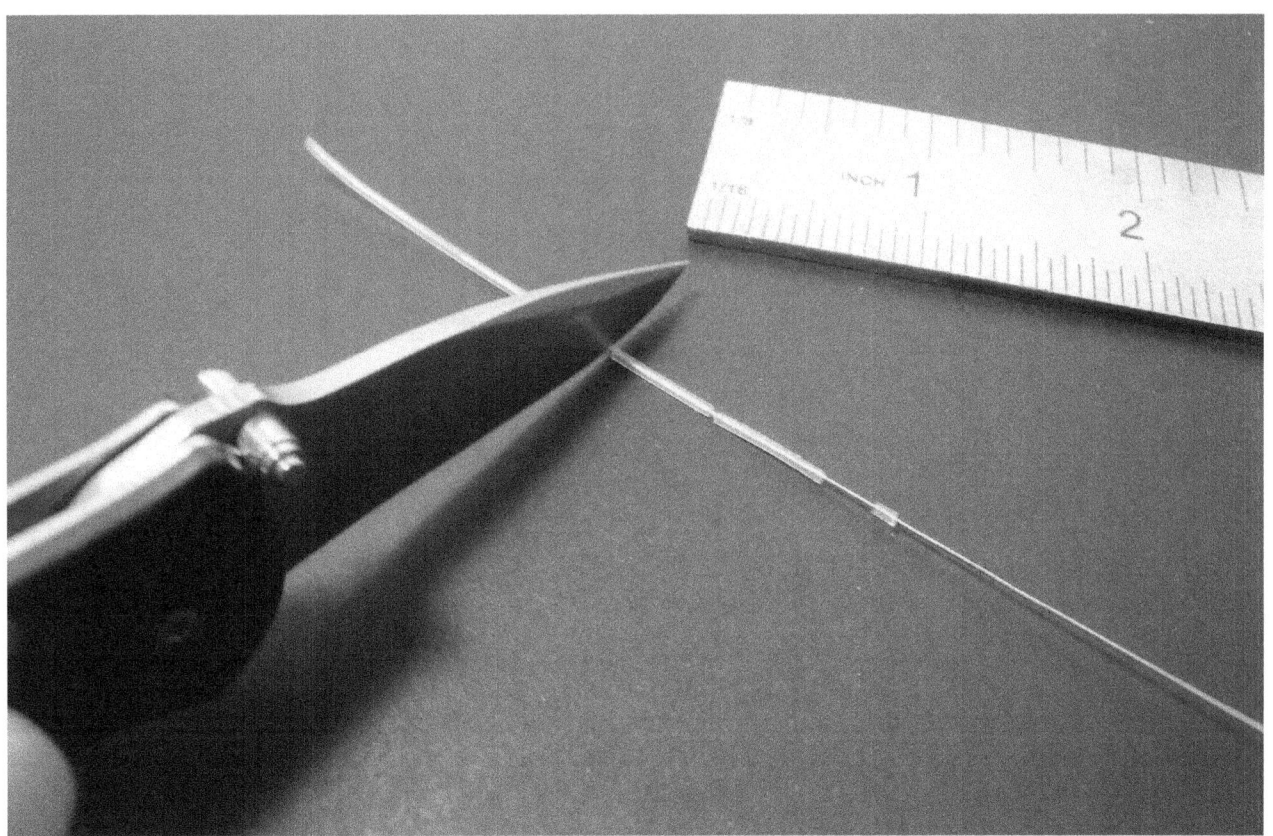

Insert a piece of music wire inside the Teflon tube before cutting. Roll the tube on a hard surface to cut around the wire. Cutting in this manner prevents the tube from deforming.

To attach the displacer gland:

1. Locate the displacer/pushrod assembled previously, the pressure chamber top plate, and the Teflon gland.
2. Slide the displacer pushrod through the top plate of the pressure chamber.
3. Set the displacer/top plate on a flat level surface with the pushrod pointing upward.
4. Slide the 7/16" (11.11 mm) gland tube onto the displacer pushrod.
5. Test for a friction free fit by verifying that the short piece of Teflon tube can fall under its own weight when dropped.
6. Let the Teflon gland drop inside the hole on the pressure chamber top plate until the end of the Teflon tube is flush with the inside surface of the top plate.
7. Apply a small bead of epoxy glue around the base of the gland tube, sealing it to the pressure chamber top plate. Take care that the glue does not touch the displacer pushrod. Allow the glue to cure.

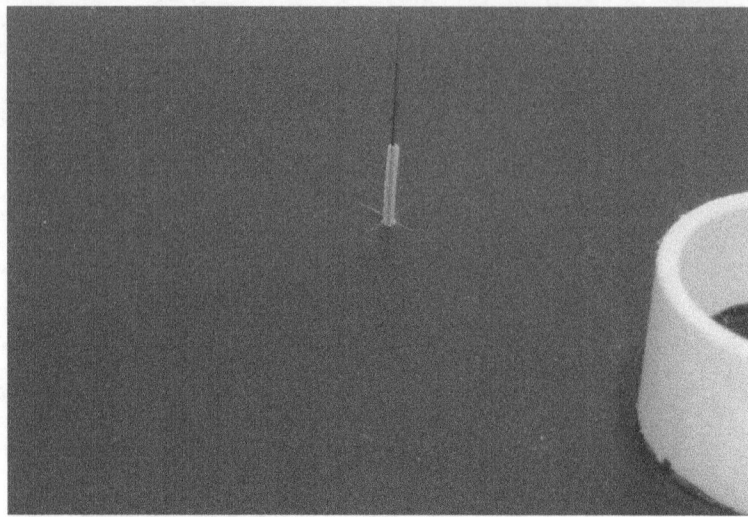

The pressure chamber top plate is placed over the displacer with the pushrod in place. The Teflon tube gland is shown here before glue is applied.

High temperature epoxy is applied around the base of the displacer gland.

Create the Flywheel

The flywheel is the largest and most visible moving part on this engine. If this part is built with care it will give the engine a very nice finished look. The goal is to make a flywheel that is perfectly round with a connection point for the axle that is squared to the flywheel and in the exact center. If the end result is less than perfect, the engine will still operate, and may even operate quite well. It just looks better if it is round, squared to the shaft, and centered.

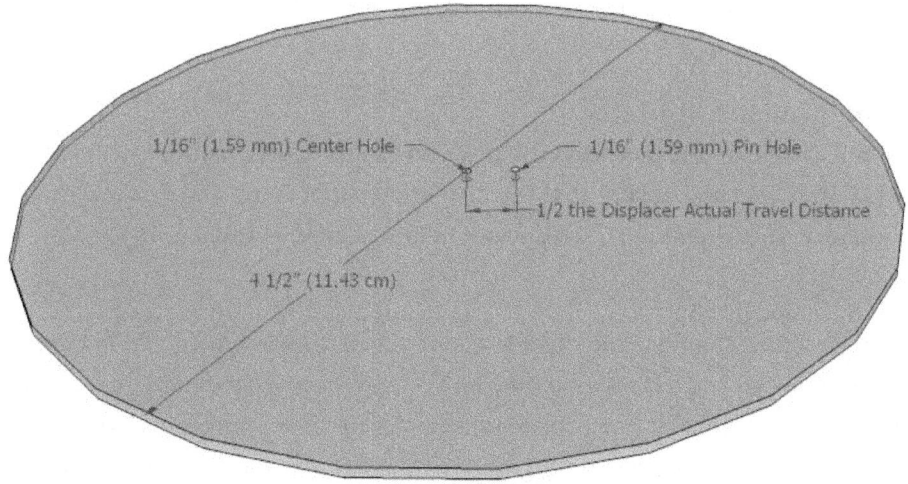

1/16" (1.59 mm) Center Hole

1/16" (1.59 mm) Pin Hole

1/2 the Displacer Actual Travel Distance

4 1/2" (11.43 cm)

The flywheel for the small engine is 4 1/2" (11.43 cm) in diameter.

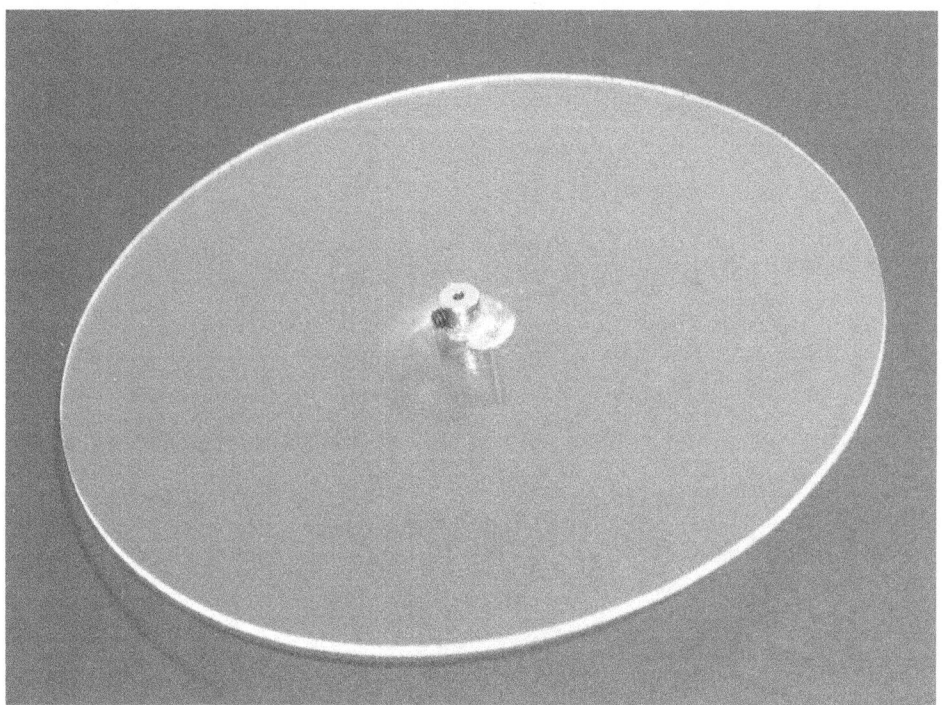

The flywheel pictured here was cut on a band saw using a simple homemade circle cutting jig. A shaft collar is attached at the center of the flywheel, and the displacer crankshaft pin will be attached on the opposite side, slightly off center.

Make a Simple Circle Cutting Jig

The process for creating a circle cutting jig and cutting flywheel is shown in detail in the instructions for the 4" Square engine in a previous chapter of this book. The process for making this flywheel is the same. Please refer to that chapter for detailed instructions for how to make the flywheel.

Using a band saw and a circle cutting jig is the recommended method for making the flywheel. If a band saw is not available, the flywheel can be cut with hand tools.

Assemble the Pressure Chamber with the Displacer

Place the displacer pushrod through the gland and place the top plate on the pressure chamber with the displacer inside. Check the fit of all the parts and ensure that the displacer can move up and down without being obstructed. Make any necessary adjustments to the displacer before the top is glued onto the pressure chamber.

Use silicone adhesive to attach the pressure chamber top plate to the round sidewall. Spread a thin layer of the adhesive on the top edge of the sidewall and smooth it with your finger. Use enough adhesive to create a seal, but no more. If too much adhesive is applied it may ooze into the inside of the pressure chamber and interfere with the motion of the displacer. Clamp the parts together with light pressure and set them aside until the silicone adhesive has cured.

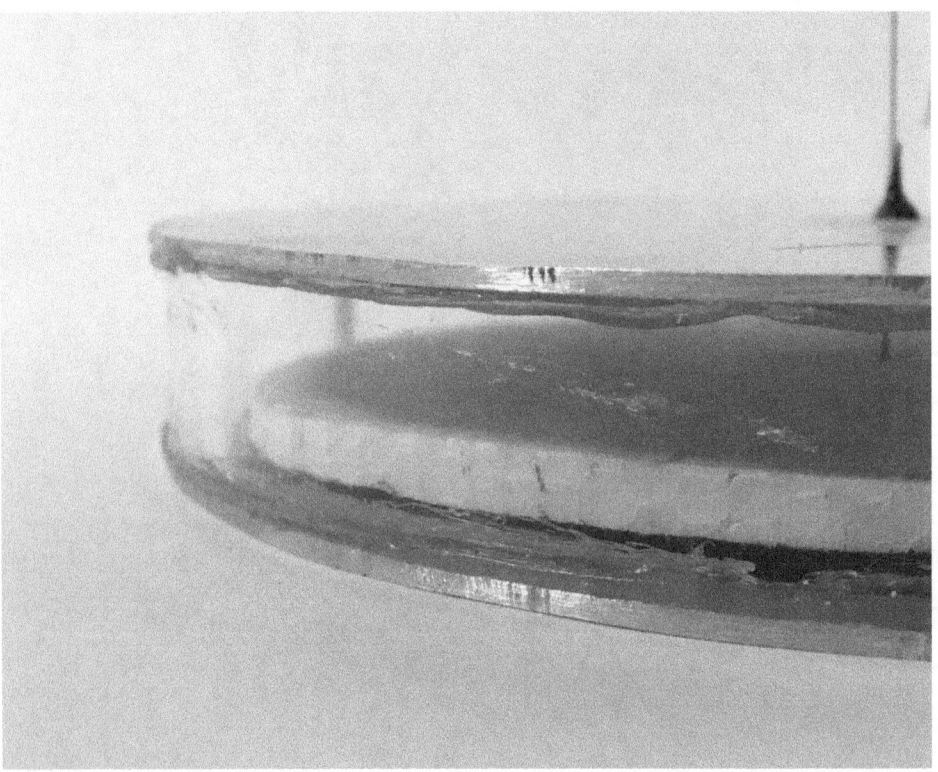

Silicone adhesive is used to attach the pressure chamber top plate because it can be opened up later if repairs are necessary.

Drill the Hole for the Crankshaft Pin in the Flywheel

The crankshaft pin is mounted in a hole that is drilled near the center hole of the flywheel. The spacing between the center flywheel hole and the crankshaft pin determines how far the displacer will travel as the flywheel rotates. The spacing between these two holes will be exactly half the _actual travel distance_ of the displacer.

154

Measure the travel of the displacer pushrod. To do this, take a measurement from the top of the pressure chamber to the top of the displacer pushrod when the displacer is on the bottom of the pressure chamber. Now lift up on the pushrod until the displacer is at the top of the pressure chamber and measure it again. Subtract the smaller number from the larger number to calculate the _total available travel distance_.

Subtract 1/16" (1.59 mm) from the _total available travel distance_ to get the _actual travel distance_. The actual travel distance will be slightly shorter than the total available distance so that the displacer will not touch the top or bottom of the pressure chamber as the engine runs. Shortening the travel distance by 1/16" (1.59 mm) will provide 1/32" (0.79 mm) of clearance above and below the displacer and prevent it from coming into contact with the pressure chamber.

It is important that the displacer does not come into contact with the pressure chamber as the engine is running. If the displacer hits the pressure chamber this will increase friction or prevent the engine from rotating freely. Also, it is good to keep a small cushion of air between the displacer and the pressure chamber top and bottom plates. The cushion of air reduces drag from what some refer to as "pull-off friction."

Pull-off friction can be demonstrated by holding a flat piece of cardboard against the ceiling with a broom handle. If the broom handle is quickly removed the cardboard does not immediately drop. The air pressure on the bottom of the cardboard holds it up until air is able to get in between the cardboard and the ceiling and equalize both pressures. This same effect can happen inside the pressure chamber if the displacer comes to rest in contact with the top or bottom plate of the pressure chamber. For this reason the engine is designed so that the displacer clearance is between 1/32" (0.79 mm) and 1/16" (1.59 mm).

Once you have determined the _actual travel distance_ of the displacer, divide that distance in half. This number will be the distance between the center hole of the flywheel and the hole for the crankshaft pin. Drill the hole for the crankshaft pin with a 1/16" (1.59 mm) drill.

The same circle cutting jig that was used to make the flywheel on the band saw is also an excellent jig for measuring and drilling the hole for the displacer crankshaft pin. The illustration shows how a pair of dividers can be used to measure the distance from the center pin to find the location for the displacer crankshaft pin hole.

Set the dividers to the distance needed for the offset of the crankshaft pin. Center one divider point on the center pin and the other point on the middle of the drill bit. Drill the hole when the alignment is correct.

Attach the Shaft Collar to the Flywheel

The flywheel is attached to the axle by means of a small round shaft collar that contains a set screw. The shaft collar is glued to one side of the flywheel in the exact center. Great care needs to be taken to align the shaft collar so that the main axle is perpendicular to the surface of the flywheel.

An alignment jig similar to what was used to attach the displacer pushrod will help attach the shaft collar with good alignment. Use a drill press to drill a 1/16" (1.59 mm) hole though a flat board. This board will serve as the alignment jig. Place a short piece of 1/16" (1.59 mm) music wire through the hole in the board. The music wire needs to be long enough to pass all the way through the board, the flywheel, and the shaft collar.

Place the flywheel over the music wire and press it flat against the surface of the board. Place the shaft collar over the music wire and press it flat against the flywheel. If the dry fit looks good, lift the shaft collar and spread a small drop of 5 minute epoxy under the shaft collar and press it against the surface of the flywheel.

The music wire will have to be removed before the epoxy is completely cured or it may become permanently attached. When the epoxy begins to harden, pull the music wire down through the alignment jig from the back side. This should leave the shaft collar glued over the center hole of the flywheel in near perfect perpendicular alignment.

156

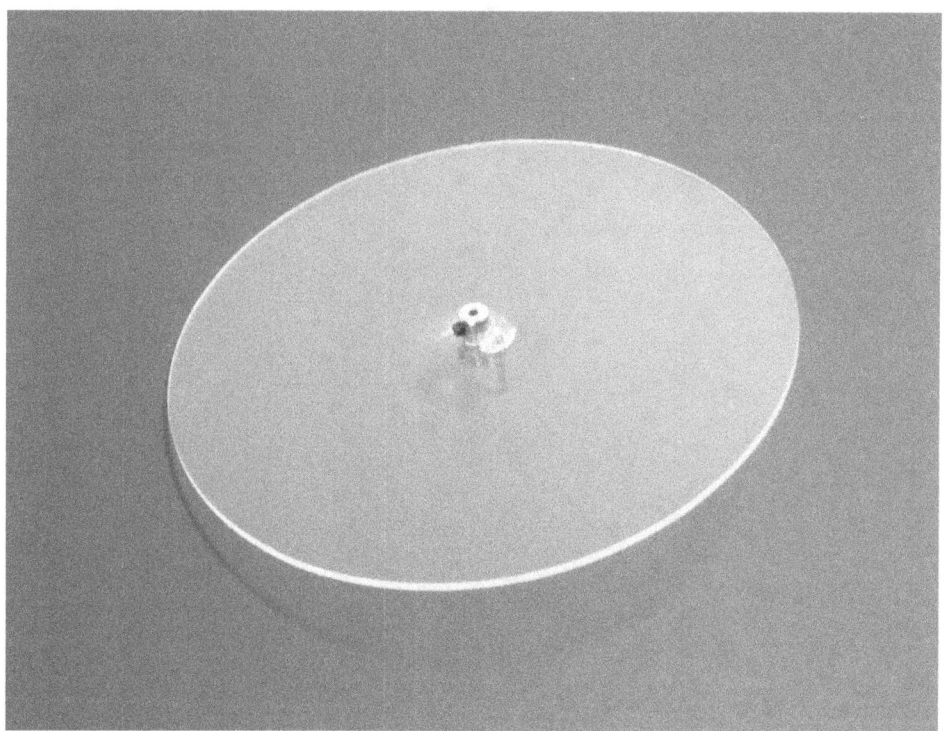

The shaft collar is attached over the center hole of the flywheel.

Make and Attach the Displacer Crankshaft Pin

There are two crankshaft pins. The displacer crankshaft pin attaches to a hole in the flywheel. The drive crankshaft pin is on the opposite end of the axle, over the drive diaphragm.

The displacer crankshaft pin is made from 0.0625" music wire, and is 1/2" (12.7 mm) long. The surface of the music wire should be smooth, with no gouges or tool marks, and it must be straight. Mark the wire where it is to be cut, clamp it in a vise, and then use the corner of a file to score the wire at the mark on two sides. Padding the vise and the pliers with paper will reduce the chances of scratching the music wire. Carefully bend the wire with pliers and it will break at the scored mark. Use a file or sandpaper to smooth the ends of the pin.

Glue the pin into the displacer pin hole on the flywheel. The location of the hole was calculated earlier. Use a small amount of epoxy to attach the pin to the flywheel. The pin must be perpendicular to the surface of the flywheel.

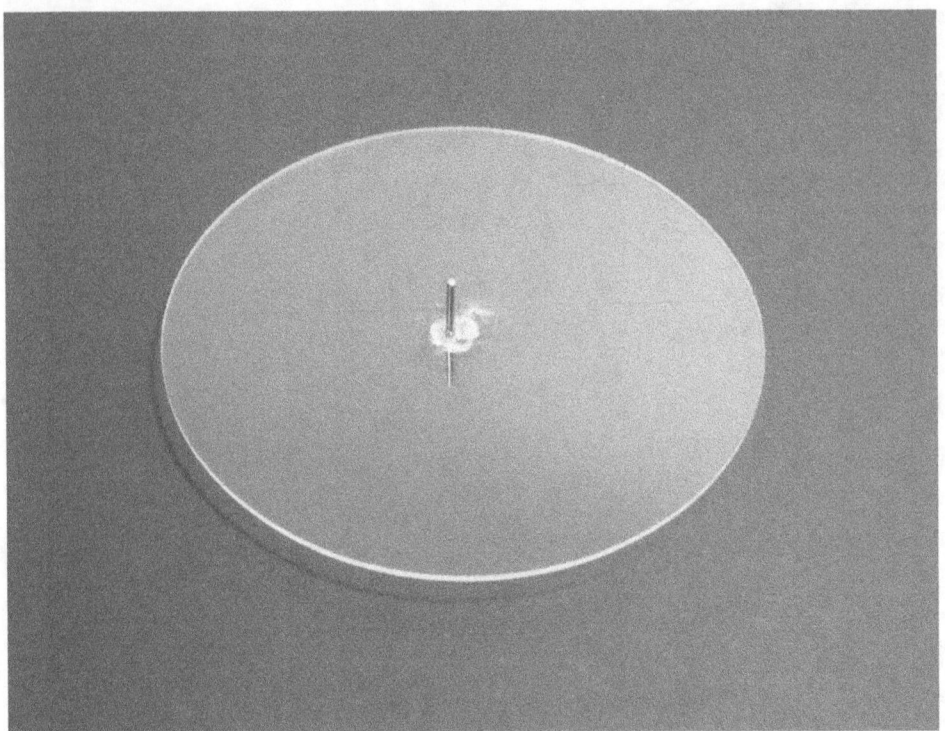

The crankshaft pin is attached to the flywheel on the opposite side as the shaft collar. The crankshaft pin is in the hole that is slightly off center.

Make the Main Axle

The axle is also made of 0.0625" music wire cut to a length of 1" (25.4 mm). Measure, score, and break the axle wire using the same technique that was used to cut the displacer crankshaft pin. Use a file or sandpaper to smooth the ends of the axle.

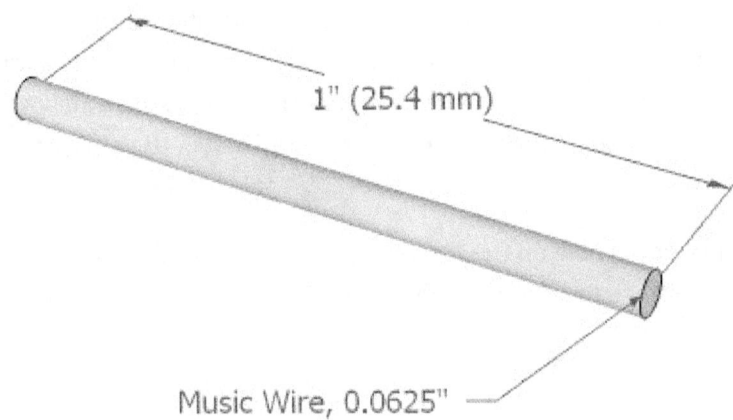

1" (25.4 mm)

Music Wire, 0.0625"

Make the Pedestal

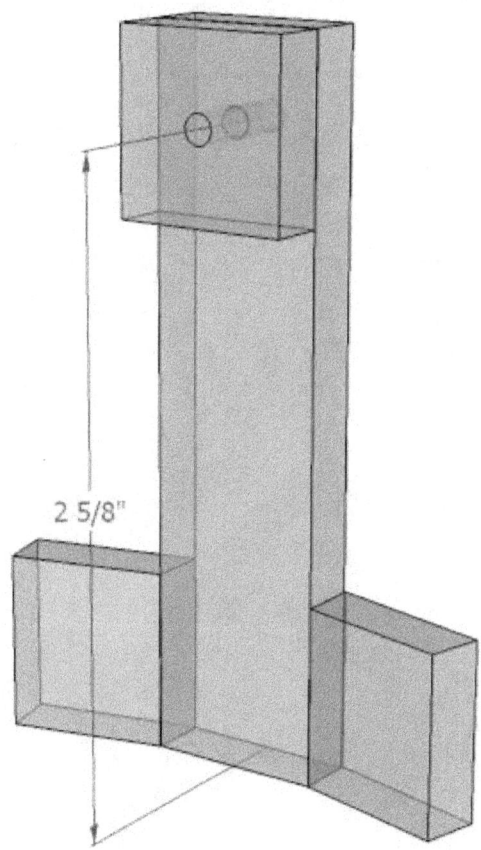

2 5/8"

The pedestal is a small post that attaches to the top of the pressure chamber. The pedestal for this engine was made from pieces of clear acrylic sheet, 1/4" (6.35 mm) thick.

This pedestal is made with a small base so that it will fit in the small space available on the top of this engine. The support pieces at the base are angled slightly to provide stability. The hole for the axle is 2 5/8" (6.67 cm) above the base of the pedestal.

The size of the hole must provide a snug fit for the Teflon tube used for the axle bushings. The tubing used in these illustrations fits snugly in a 7/64" (2.78 mm) hole. Measure to verify the hole size needed for the bushing material that will be used.

This picture shows the acrylic pedestal parts before assembly. Note the angled surfaces on the support pieces at the base.

Cut the acrylic parts for the pedestal. The pedestal can be assembled using clear five minute epoxy.

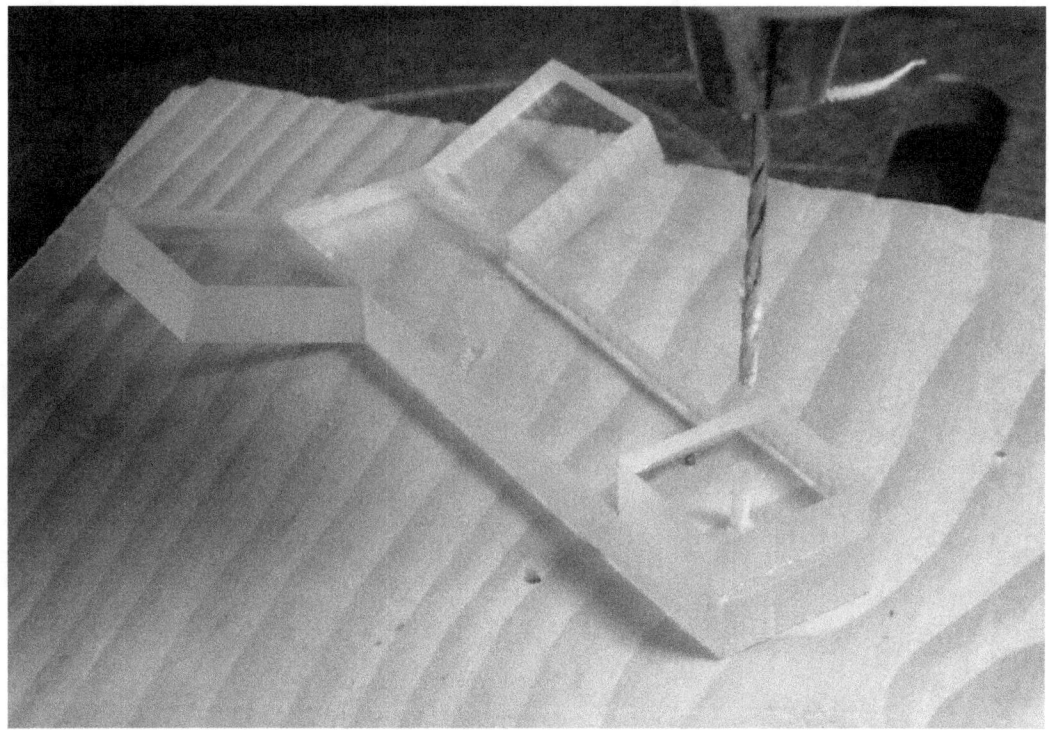

This picture shows the pedestal after the parts have been glued together. Drill a hole for the axle bushing. Select a drill bit that will provide a snug fit for the Teflon tube that will be used as the axle bushing.

Cut and Mount Teflon Tube Axle Bushings

The axle bushings are made from AWG-14 (0.066") Heavy Wall Extruded Teflon Tube. Two pieces are required. Each piece is 3/16" (4.76 mm) long.

Cut the tubing by first placing a piece of 0.0625" music wire inside the tube at the place to be cut. Place the tubing on a flat surface and roll the tubing while cutting with a sharp knife or razor blade. Cut down to the music wire as the tubing is rolled on the flat surface. Cutting in this manner prevents the tubing from being deformed during the cutting process. Cutting the tubing without the shaft inside can cause the tubing to flatten or kink at the point of the cut, causing friction in the bushing.

Insert the small pieces of Teflon tube into each end of the hole in the pedestal. Leave a small amount of tubing (about 1/16" (1.59 mm) or less) protruding from the hole on both sides. No gluing will be required if the hole is the correct size.

The outside diameter of the axle is 0.0625", which allows it to turn freely inside the 0.066" Teflon tube. Insert a piece of axle material into the bushings to test the fit. Make any adjustments necessary to enable the axle to spin freely in the bushings.

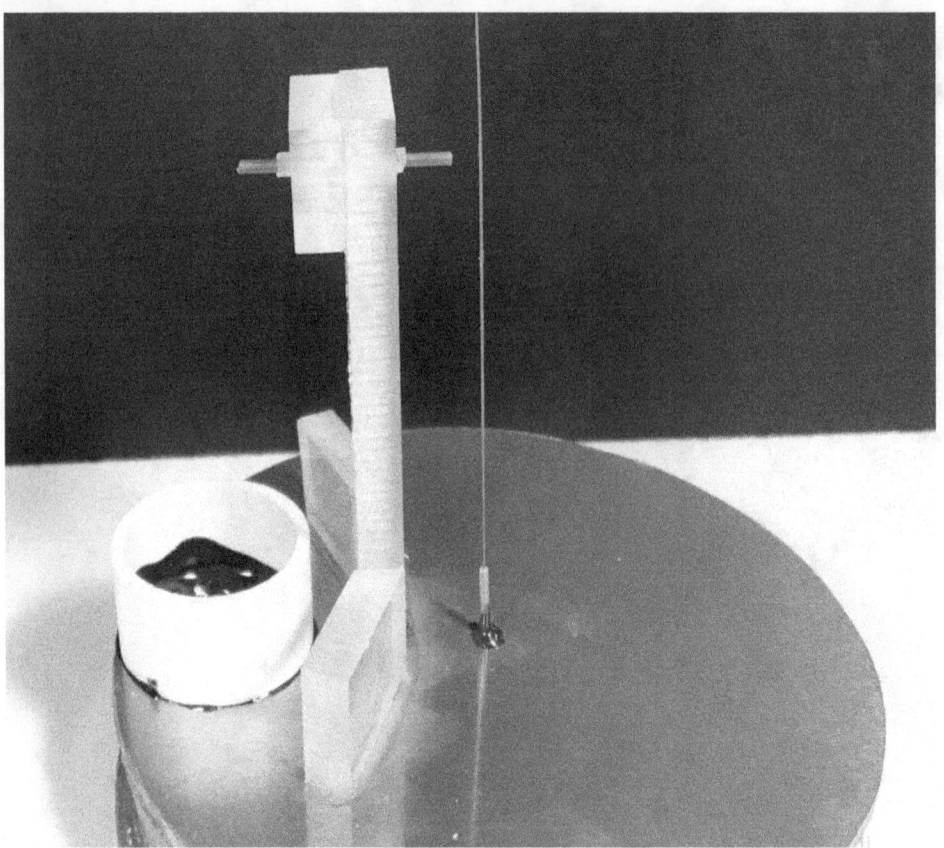

The pedestal is shown here with the Teflon bushings installed.

Dry-Assemble and Measure for Locating the Pedestal

Attach the axle to the flywheel using the shaft collar, and insert this assembly into the bushings of the pedestal. The flywheel will be on the straight side of the pedestal.

Position the pedestal on top the pressure chamber between the drive cylinder and the displacer pushrod. The axle must be centered over the displacer pushrod and the drive cylinder. The pedestal must be positioned so that both the pushrods can be attached to their respective pins at a 90^0 angle to the axle. Once the ideal position for the pedestal has been found, mark the position with a pencil. Set the pedestal aside for now. It will be attached after the drive crankshaft has been assembled.

Make the Drive Crankshaft

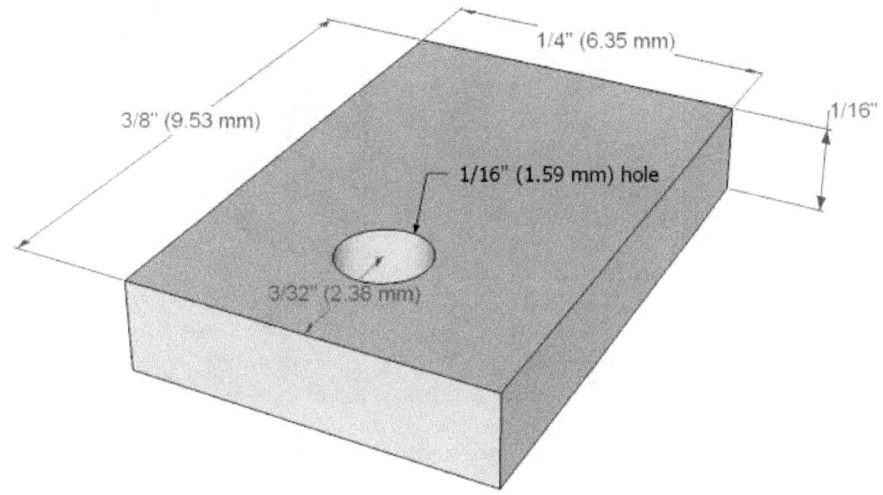

The drive crankshaft plate creates the offset for the crankshaft that attaches to the drive diaphragm. It holds the drive crankshaft pin parallel to the axle with an offset of 3/16" (4.76 mm). A hole is drilled to attach the crankshaft pin to one side. A shaft collar is glued to the opposite side and is the attachment point for the main axle. The hole for the pin is 3/16" (4.76 mm) from the center of the plate.

Draw the shape of the crankshaft plate on a piece of aluminum stock and drill the hole before cutting the plate. The plate is made from 0.062" aluminum sheet, the same material recommended for the top and bottom plates of the pressure chamber. The dimensions of the plate are 1/4" (6.35 mm) wide by 3/8" (9.53 mm) long. Drill a 1/16" (1.59 mm) hole at a position 3/32" (2.38 mm) from one end of the plate. Cut the small plate from the aluminum sheet after the hole has been drilled.

Cut the drive crankshaft pin from a piece of 0.062" music wire. The length of the pin is 1/2" (12.7 mm). Attach the pin to the hole in the plate with epoxy. The pin must be perpendicular to the surface of the plate. The pin is mounted on the front side of the plate.

Place a mark at the center of the back side of the plate, 3/16" (4.76 mm) away from the position of the pin. This mark will be used to position the shaft collar to the back side of the plate. Use epoxy to attach a 1/16" (1.59 mm) shaft collar to the back side of the plate. Take care that the set screw of the shaft collar does not become fouled with epoxy.

The pictures show multiple crankshaft plates. Only one is required. Making several crankshaft plates with different offset measurements makes it possible to change the travel distance of the drive mechanism. This can be a useful adjustment for fine tuning the engine to operate in different temperature environments.

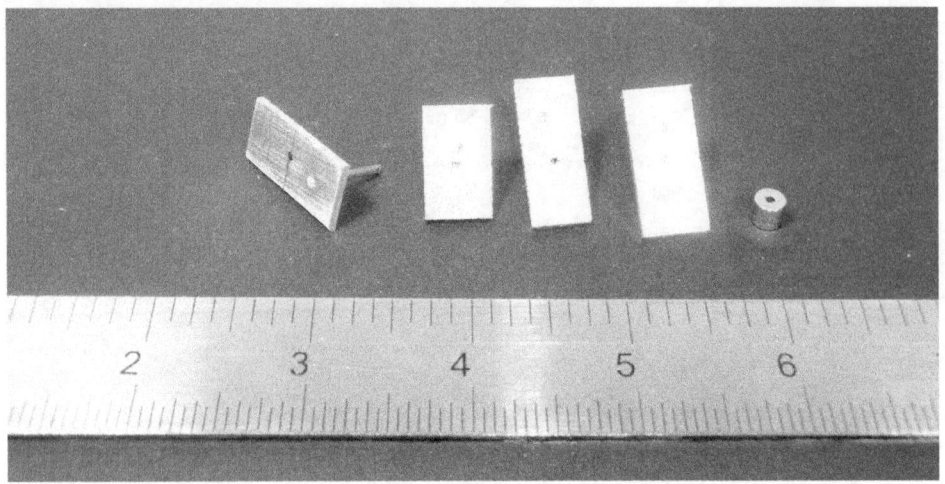

The crankshaft plate and the shaft collar are ready to be joined together with epoxy.

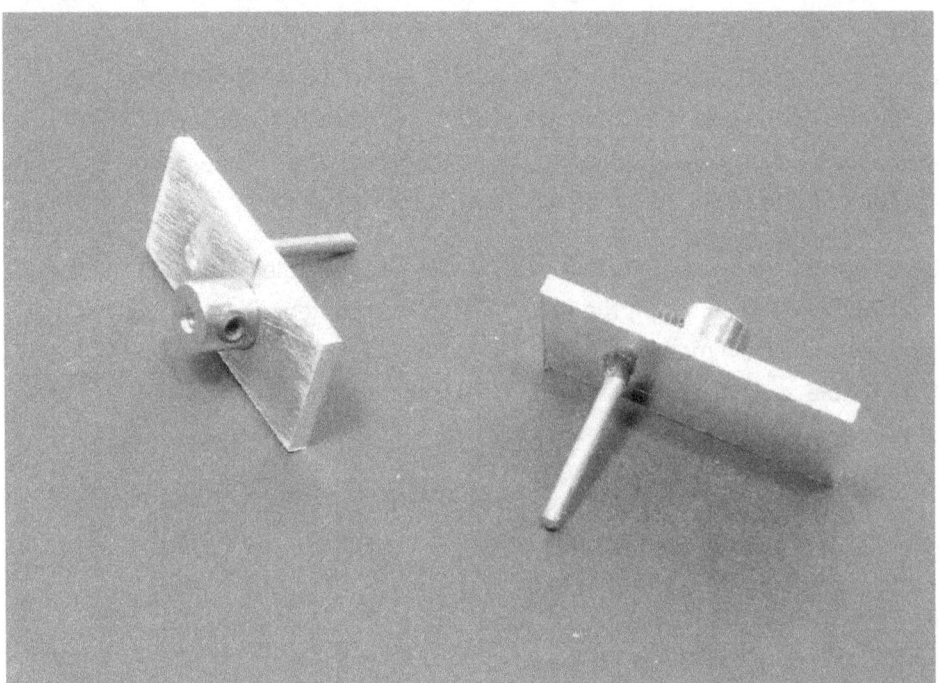

The crankshaft pin, plate, and shaft collar have been assembled.

Attach the Pedestal to the Pressure Chamber Top

The pedestal needs to be mounted so that it is aligned well with the displacer pushrod and the drive cylinder. Temporarily place a piece of straight music wire through the axle bushings to visually confirm the alignment. This will make is easier to visually check the alignment and confirm the previous marks. Glue the pedestal in place with epoxy once the correct position is confirmed.

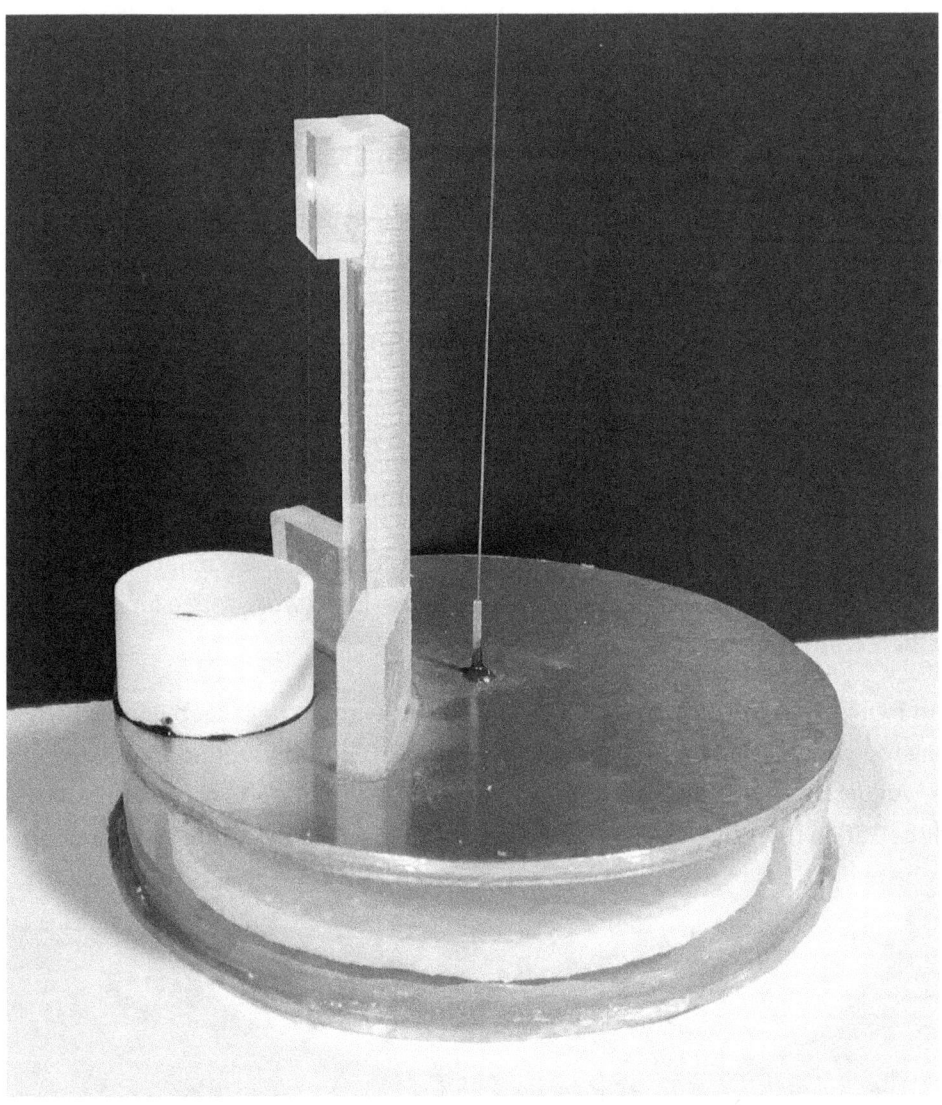

The pedestal is attached to the pressure chamber top. There is enough room for the flywheel between the pedestal and the displacer pushrod, and the main axle is aligned with the center of the drive cylinder and the displacer pushrod.

Test the Travel Distance of the Displacer

Install the axle, flywheel, and drive crankshaft on the pedestal. Position the flywheel so that the displacer crank pin is at the bottom position of its rotation. Use a marker or a piece of tape to make a reference mark on the displacer pushrod at the point where it comes in contact with the crank pin. Now rotate the flywheel until the pin is at the top of its rotation and raise the displacer pushrod until the mark is once again even with the pin. There should be about 1/16" (1.59 mm) free space between the displacer panel and the top of the pressure chamber when the displacer crank pin is at the top of its rotation. If it appears that the displacer will be able to move up and down without impacting the top or bottom of the pressure chamber, proceed to the next step. If the displacer crank is moving the displacer too far and it is impacting the engine, correct the problem by relocating the crank pin closer to the center of the flywheel.

Trim the Displacer Pushrod
Allow the displacer panel to rest on the bottom of the pressure chamber. Measure up from the top of the pressure chamber 1 1/2" (38.1 mm) and trim the displacer pushrod at this point.

Create the Displacer Connecting Rod and Teflon Bushing
The displacer connecting rod completes the connection between the top of the displacer pushrod and the displacer crank pin on the flywheel. A flexible connection is made to the displacer pushrod with two thin pieces of duct tape. The top end of the connecting rod is a piece of Teflon tube that will slip over the displacer crank pin on the flywheel. The displacer connecting rod is made from 0.015" music wire, which is the same size as the displacer pushrod.

Cut a piece of AWG-14 heavy wall extruded Teflon tube to a length of 7/16" (11.11 mm). Use the same cutting method described earlier so that the tubing does not become deformed at the cut.

Cut a piece of 0.015" music wire to a length of at least 3 1/2" (88.9 mm) to construct the connecting rod. The exact length is not critical because it will be trimmed to fit. It may be easier to work with a longer piece and trim it to length after the bends are completed.

Place the short piece of Teflon tubing on a piece of axle stock (0.062" music wire) to help keep it straight while wrapping the connecting rod wire around the tubing. Make at least 1 1/2 wraps around the tubing with the connecting rod wire. Adjust the connecting rod wire so that it is at a 90° angle to the Teflon tube. The connecting rod wire and the Teflon tube should form the shape of a "T." Make the wraps tight enough so that they grip the Teflon tube and it does not fall out. Trim any excess wire when finished.

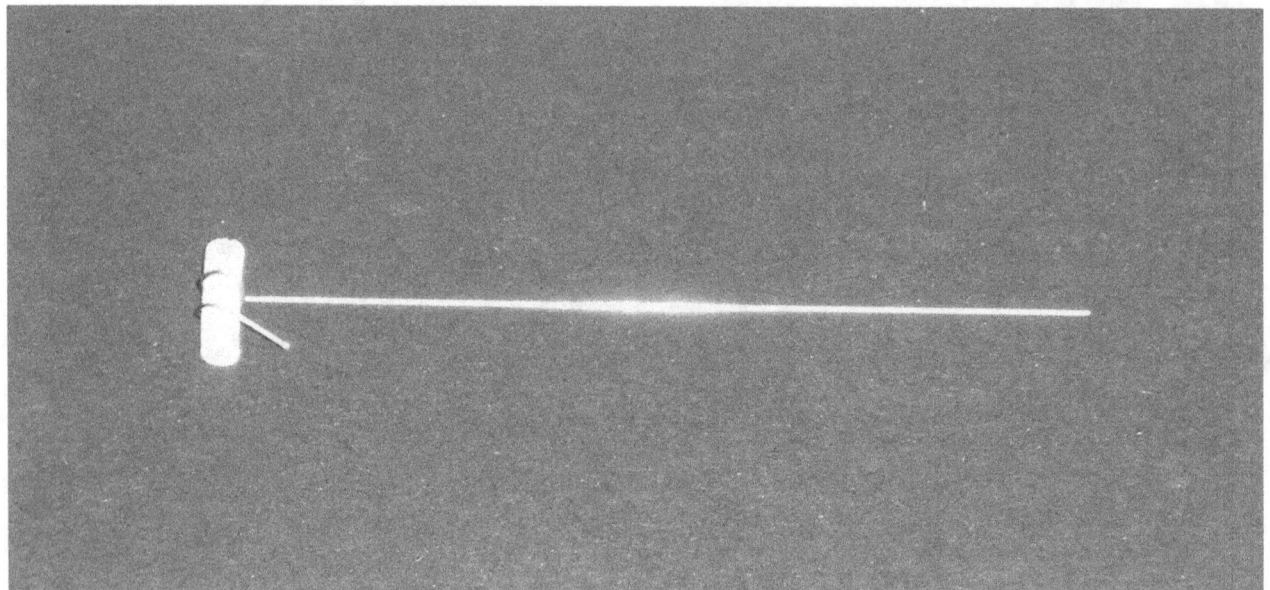

The Displacer Connecting Rod is fashioned from a length of music wire that is wrapped around a short piece of Teflon tube.

Attach the Displacer Connecting Rod

The connecting rod can now be installed between the flywheel and the displacer pushrod. Place the Teflon tube over the drive crankshaft pin on the flywheel. With the displacer panel resting on the bottom of the pressure chamber and the flywheel pin in its lowest position, trim the length of the connecting rod so that there is a gap between the end of the displacer pushrod and the end of the connecting rod. The gap should be between 1/32" (0.79 mm) and 1/16" (1.59 mm).

Cut two small pieces of duct tape to 1/8" (3.18 mm) x 1/2" (12.7 mm). Lift the displacer so that the end of the displacer pushrod is near the end of the connecting rod and fasten the two pieces together with the small pieces of duct tape. Position the duct tape so that the joint bends correctly in order to accommodate the motion of the flywheel.

Rotate the flywheel and observe the motion of the displacer panel. It should travel up and down inside the pressure chamber without touching the top or the bottom of the pressure chamber. Make any adjustments necessary so that the displacer does not come into contact with the top or bottom of the pressure chamber.

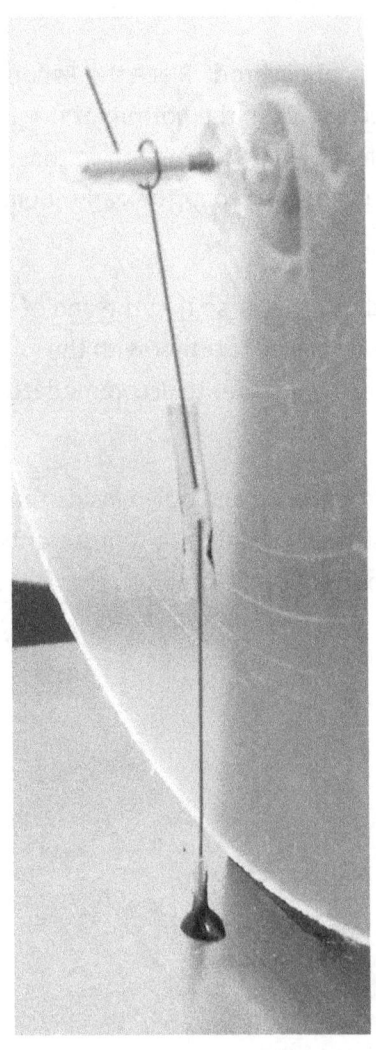

In this picture, the displacer connecting rod has been trimmed to leave a small gap between the ends of the two rods. One piece of duct tape has been applied to the joint. Note the direction of the bend at the joint and how the alignment of the tape allows it to act as a hinge. Apply a second piece of tape when proper positioning of the parts has been confirmed.

Create the Drive Pushrod and Teflon Bushing

The drive pushrod is made just like the displacer connecting rod was made, except there is no duct tape joint in the middle of the shaft. There is a Teflon tube at the top end of the pushrod. The Teflon tube rides on the crankshaft pin. The bottom end is bent into a loop that is folded over so that it mounts flat against the drive diaphragm.

Cut a piece of AWG-14 Teflon tube to a length of 7/16" (11.11 mm). Insert a piece of 0.062" music wire inside to help hold it while wrapping a piece of 0.015" music wire around it. Make at least 1 1/2 turns around the tubing, as before. Adjust the tubing so that it is held snugly at a 90^0 angle to the pushrod.

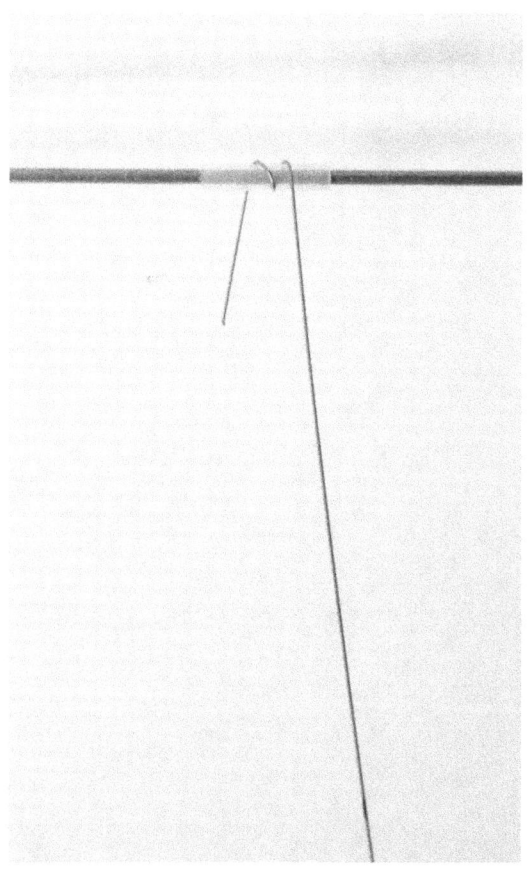

In this picture, the Teflon tube has been placed on a piece of axle wire and the pushrod wire is wrapped around the tubing to attach it to the pushrod.

Measure the distance between the center of the axle and the top of the drive cylinder. This will be the finished length of the pushrod.

Make a 90° bend in the pushrod using the measurement just obtained. The distance from the axle to the top of the drive cylinder should be the same as the distance from the Teflon tube to the 90° bend.

This picture shows the 90° bend aligned with the top of the drive cylinder.

Use needle nose pliers to make a loop at the bottom of the pushrod. The loop should be about 7/16" (11.11 mm) in diameter. The loop is made in such a way so that if the connecting rod was placed on a flat surface it could stand upright with the loop flat against the table top.

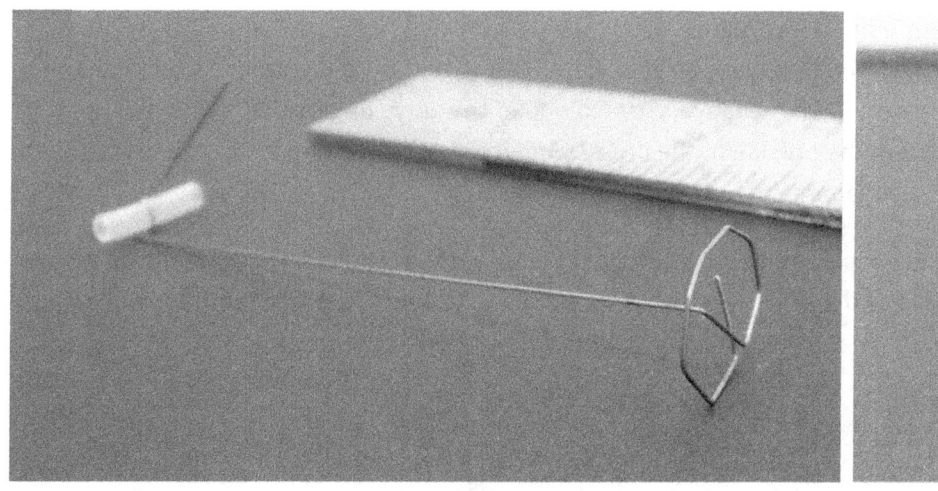
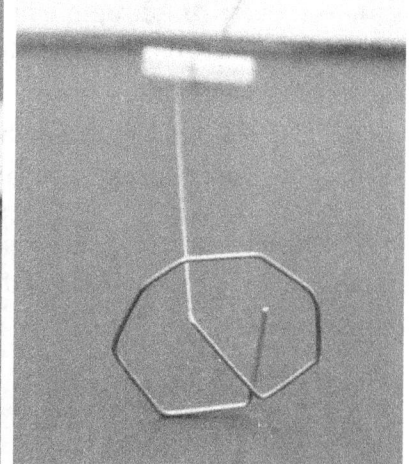

These pictures show the loop formed at the 90° bend in the pushrod.

Verify the fit of the pushrod. Place it on the drive crankshaft pin and rotate the flywheel. The loop on the bottom should rise above the top of the drive cylinder at its highest point, and it should drop down into the drive cylinder at its lowest point. It should be close to even with the top of the drive cylinder when at its midpoint between high and low.

If getting the correct length is a challenge, consider making the shaft slightly longer than required, and then place a Z shaped bend near the middle of the pushrod. The Z bend can be manipulated to fine tune the length of the pushrod after it is installed.

Create the Drive Diaphragm from a Latex Glove

The drive diaphragm is made from the fingertip of a latex glove. Wash and dry a latex glove so that all the powder is removed. Cut the fingers from the glove. Stretch one of the glove fingers over the top of the drive cylinder. Pull the latex down the outside of the drive cylinder until there is only a small amount of slack near the center of the drive diaphragm.

The drive diaphragm should stay in place without any help. If it appears to be slipping, secure it by placing a rubber band around the outside of the drive cylinder. A rubber band can be made from another glove finger if necessary.

The drive diaphragm is made from the finger tip of a latex glove. There must me a small amount of slack to the diaphragm when it is installed.

Attach the Drive Pushrod to the Drive Diaphragm

The loop at the end of the drive pushrod should rest flat against the drive diaphragm and there should be no sharp wires threatening to puncture the diaphragm. Hold the loop against the center of the diaphragm and attach it using Superglue®.

This picture shows the drive pushrod attachment to the drive diaphragm. The wire pushrod is attached to the latex diaphragm with Superglue®. The diaphragm is made from the finger tip that has been cut from a latex glove.

Set the Crankshaft Timing Angle

Adjust the flywheel and the drive crankshaft so that there is a 90^0 offset between the displacer crank pin and the drive crank pin. The direction of the offset will determine the direction the motor rotates when running. The drive mechanism will follow 90^0 behind the motion of the displacer mechanism. That means when the displacer is all the way up at the top of the rotation, the drive mechanism will be halfway up. When the displacer is all the way down, the drive mechanism will be halfway down. Use the set screws on the shaft collars to hold the flywheel and the drive crankshaft in place.

Adjust the Drive Diaphragm Tension

The drive diaphragm should be adjusted so that there is just enough slack in the latex to allow the engine to rotate without stretching the material. If the material is too tight the engine will not run well because extra energy will be required to stretch the diaphragm. If the diaphragm is too loose the engine will not run well because the loose diaphragm will inflate and deflate without causing the crankshaft to move. Adjusting the tension of the drive diaphragm is one of the adjustments that can be made to fine tune the performance of the motor.

Check all the Connections

The engine should now be fully assembled and ready for its first run. Check all the connections by rotating the flywheel slowly and observing all the moving parts. Nothing should be falling apart when the flywheel is rotated. If the connecting rod or the pushrod becomes disconnected during this test, make adjustments so that they do not fall off.

Observe the motion of the displacer panel inside the pressure chamber and ensure that it does not impact the top or the bottom of the pressure chamber. It should move without any obstruction.

Run the Engine!

This Stirling Engine should run well over hot water. Fill a coffee cup or similar container with near-boiling water. Place the pressure chamber on top the cup of hot water. Allow it to warm up for 10 to 20 seconds. Turn the flywheel to start the engine.

The motor will continue to run as long as there is a temperature differential of 20° F (11° C) (or more) between the top and bottom surfaces of the pressure chamber. It may be possible to fine-tune the engine to operate on an even lower temperature differential with a little care and patience.

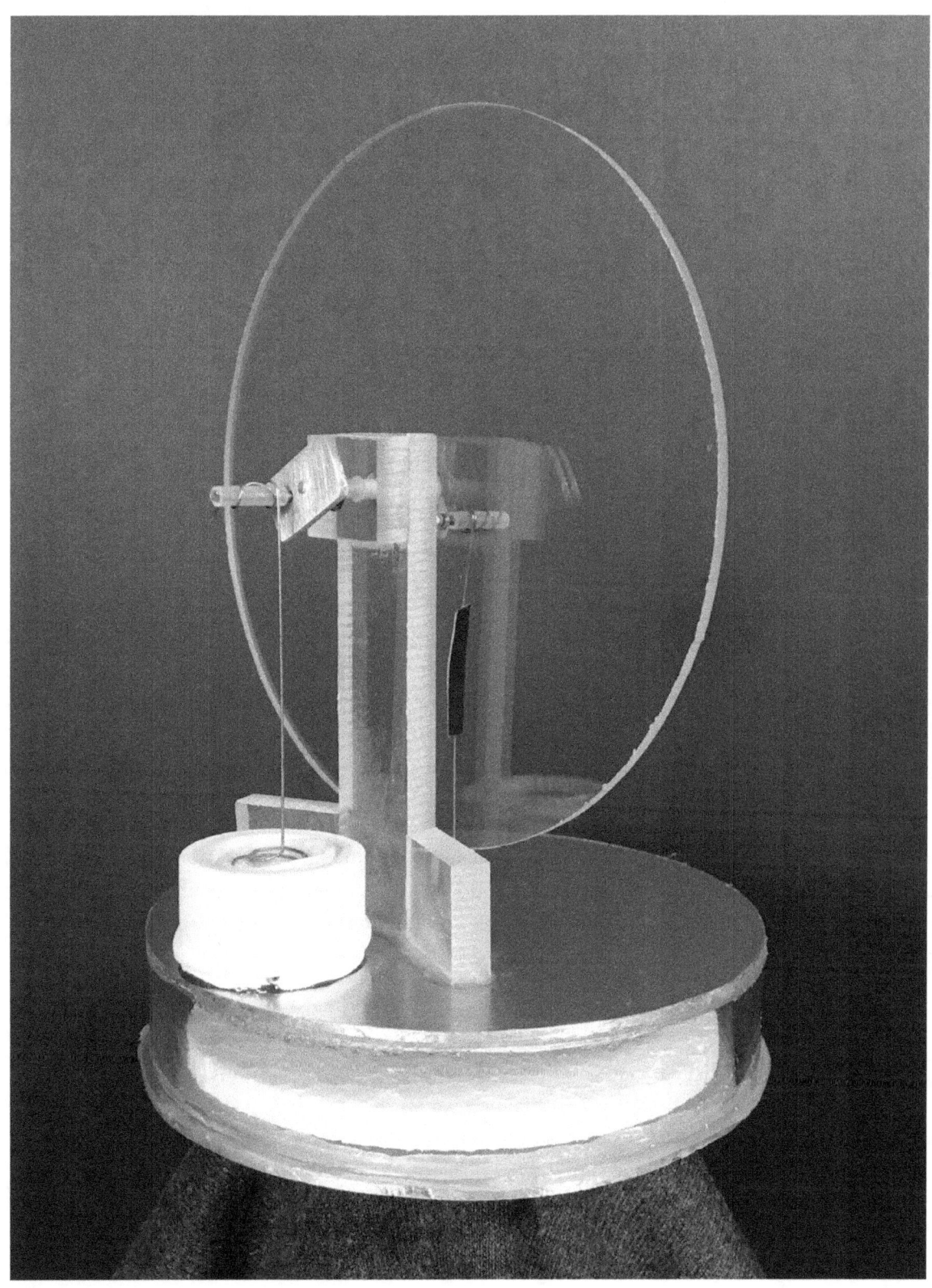

The finished 4" (10.2 cm) round engine.

Trouble Shooting Tips

If the engine is not running well, it may be because of a problem in one of these four areas:

Temperature Differential: As mentioned previously, the engine should run with a temperature differential of 20° F (11° C). Setting the engine on a cup of near-boiling water in a 70° F (21° C) room should provide a temperature differential of about 100° F (56° C). If the engine is not running under these conditions, there are one or more other problems that will need to be fixed, such as a small pressure leak or friction.

It may be possible to overcome a small pressure leak or friction by increasing the temperature differential. This will increase the power output of the engine and possibly overcome a small amount of friction or a small pressure leak.

Increase the temperature differential by adding ice to the top of the motor while the bottom is being warmed by the heat source. Do not attempt to add more heat, as this can damage the engine. The Styrofoam displacer material may melt if the heat source is too hot.

Pressure Leaks: It does not take much of a leak to prevent the engine from running well. There are a couple of ways to test for a pressure leak. The first method is to observe the behavior of the diaphragm when the engine is at running temperature.

Disconnect the drive diaphragm pushrod from the crank pin. Place the engine on a cup of hot water and wait a few moments for the bottom side to heat up. Now, rotate the flywheel so that the displacer rises and falls inside the pressure chamber and observe the motion of the drive diaphragm. The diaphragm should move up and down in response to the heating and cooling of the air inside the pressure chamber. If this motion is not present, or if it is very limited, there may be a pressure leak. It may also be possible that the tension of the diaphragm is too tight or too loose.

The other method for leak testing also involves removing the diaphragm pushrod from the crank pin. Once it is disconnected, pull upward on the pushrod for 5 to 10 seconds to inflate the diaphragm. Release the pushrod and observe the diaphragm. If it immediately deflates and returns to a low or neutral position, there may be a pressure leak.

Pressure leaks can happen at a number of places:

- Holes in the drive diaphragm
- Leaking around the edge of the drive diaphragm
- Leaking glue joints in the pressure chamber
- Leaking through excessive clearance around the displacer pushrod.

It may take a bit of detective work to find the leak. It may be possible to patch a small leak with a small drop of glue or silicone sealant, or it may be necessary to replace the defective part.

Friction: Small amounts of friction can have a huge impact on an LTD Stirling engine's ability to run. Friction occurs at every point where two moving parts touch, and at every point where a moving part contacts the atmosphere. In the micro-horsepower world of LTD Stirling engines, a tiny bit of friction can stop the engine from performing.

Check the rotation of the axle by removing the connections to the displacer and the drive diaphragm and spinning the flywheel. The flywheel should coast to a stop after about 30 seconds after receiving a good spin by hand. If the flywheel does not spin freely there is a problem with friction somewhere in the axle assembly. Locate the cause of the friction and repair the problem.

The flywheel rotation should be smooth and silent during the spin test. Vibration and noise are both indications of friction.

Crankshaft Timing: There must be a 90^0 phase difference between the two crank pins. This means that when one pin is in the 12:00 o'clock position, the other one is at either 9:00 o'clock or at 3:00 o'clock. The engine will run with a phase difference in either direction. The only difference will be the direction the engine rotates while running.

When the bottom of the engine is heated, the engine will run with the motion of the displacer moving ahead of the drive diaphragm. This means that when the displacer is at the top of the pressure chamber, the drive diaphragm is halfway up and moving in an upwards direction. As the flywheel rotates and the displacer comes to the lowest point in its travel, the drive diaphragm is halfway down and moving downward. It is important to know which way the engine will run so that the initial push to get it started is in the same direction.

Engine #7: Large Round Engine 6" (15.2 cm)

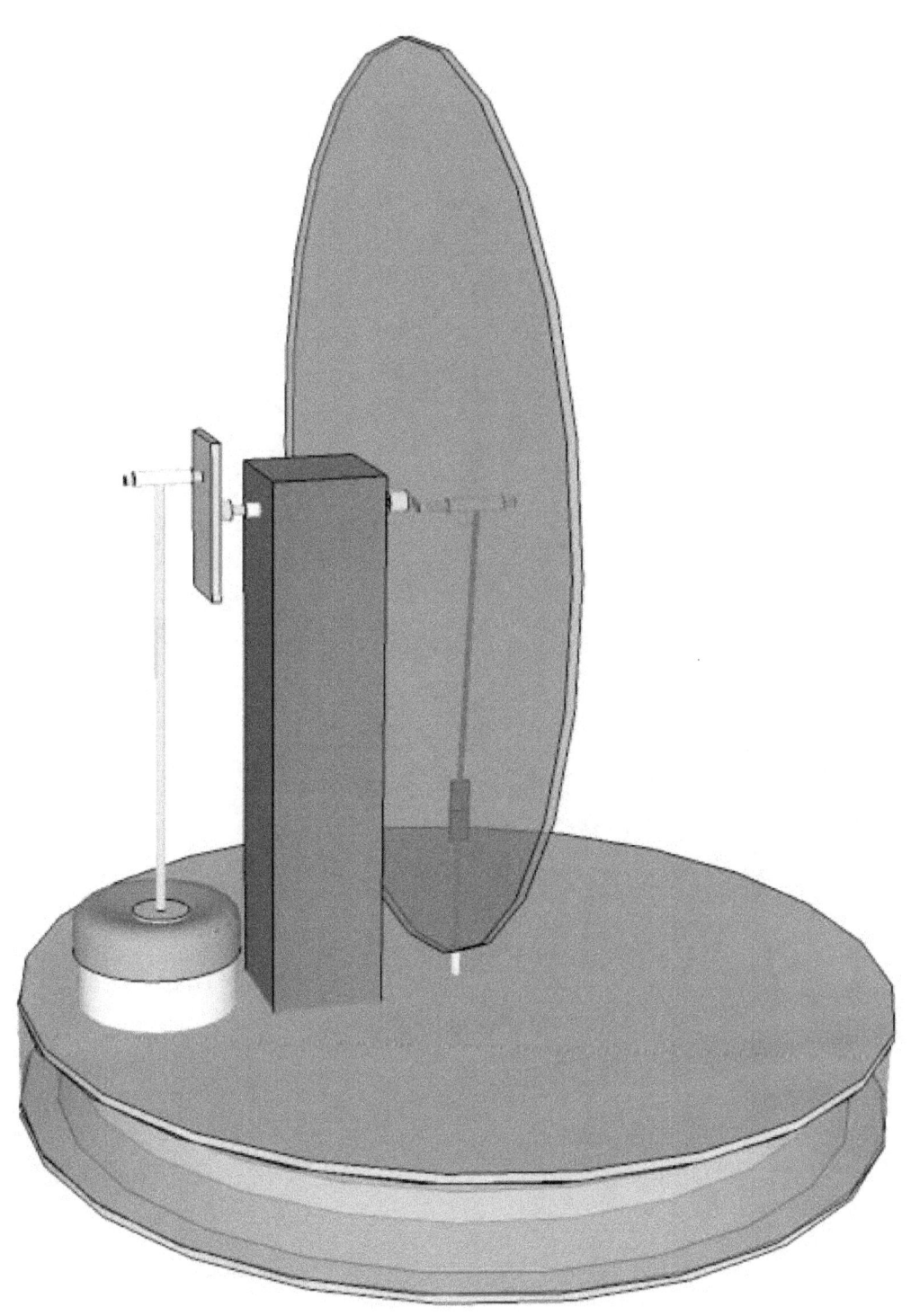

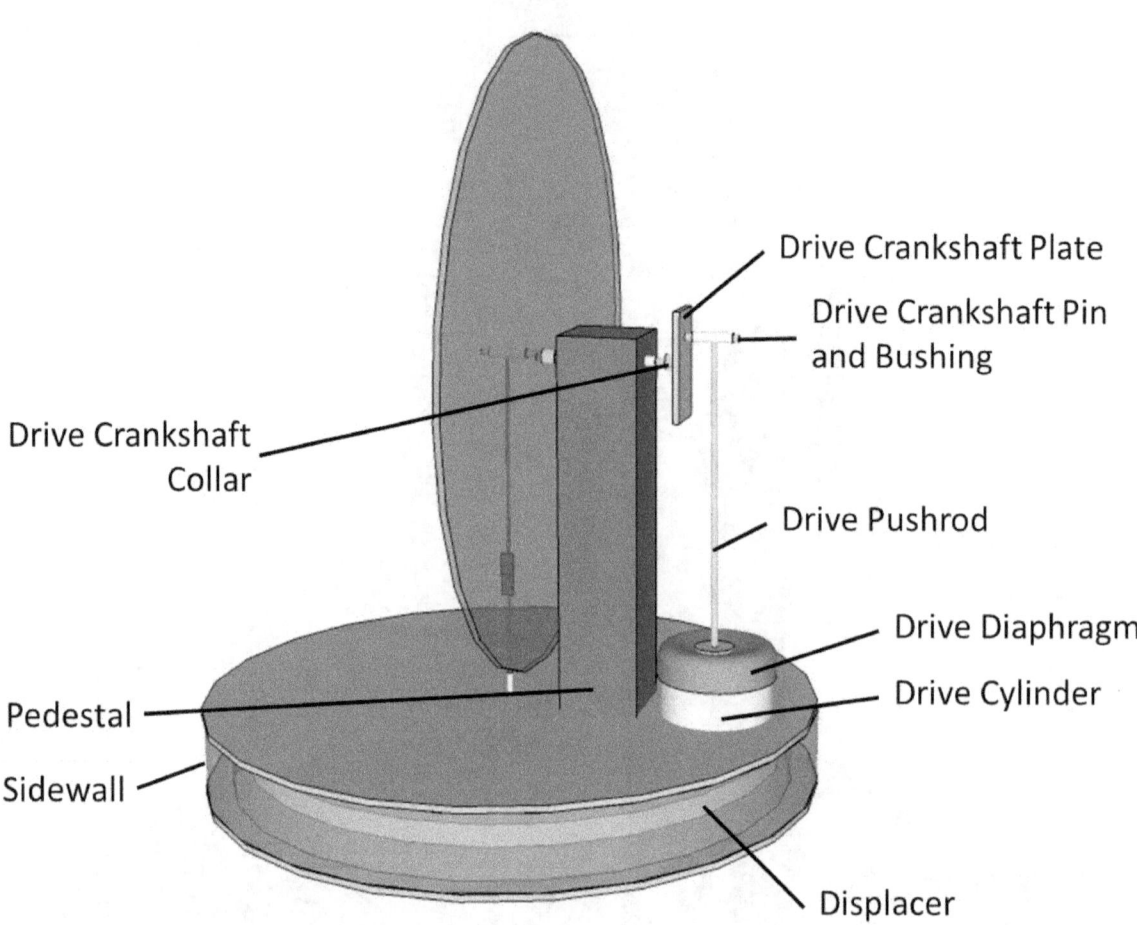

Drive Crankshaft Plate

Drive Crankshaft Pin
and Bushing

Drive Crankshaft
Collar

Drive Pushrod

Drive Diaphragm

Drive Cylinder

Pedestal

Sidewall

Displacer

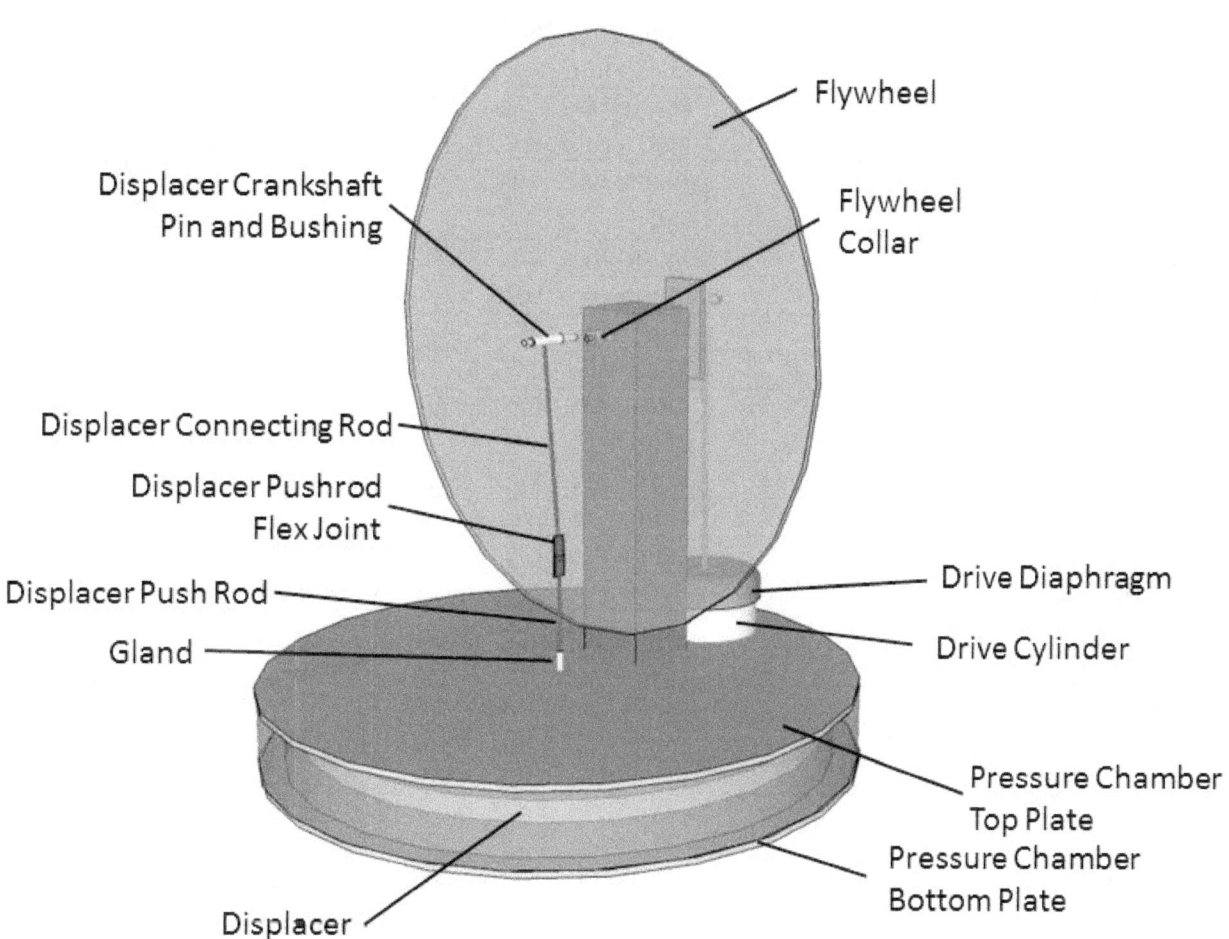

Flywheel

Displacer Crankshaft
Pin and Bushing

Flywheel
Collar

Displacer Connecting Rod

Displacer Pushrod
Flex Joint

Displacer Push Rod

Gland

Drive Diaphragm

Drive Cylinder

Pressure Chamber
Top Plate
Pressure Chamber
Bottom Plate

Displacer

Parts and Materials

Part #	Part Name	Description
01	Flywheel	Clear Acrylic Sheet, round, 0.06" x 5 3/4" (14.61 cm) diameter*
02	Flywheel Collar	Shaft Collar, 1/16" (1.59 mm)
03	Pressure Chamber Bottom Plate	Aluminum Sheet, round, 0.062" x 6" (15.24 cm) diameter
04	Pressure Chamber Top Plate	Aluminum Sheet, round, 0.062" x 6" (15.24 cm) diameter
05	Displacer	Styrofoam, round, 0.20" (5 mm) x 5 1/4" (13.34 cm) diameter
06	Displacer Crankshaft Pin	Music Wire, 0.0625 x 1/2" (12.7 mm)
07	Displacer Crankshaft Bushing	AWG-14 (0.066") Heavy Wall Extruded Teflon Tube, 3/8" (9.53 mm)
08	Pressure Chamber Sidewall	Formed Clear Acrylic, 1/4" (6.35 mm)** x 11/16" (17.46 mm) x approx. 20" (50.8 cm)
09	Pedestal	Wood, 3/4" (19.05 mm) x 3/4" (19.05 mm) x 3 1/2" (8.89 cm)
10	Displacer Connecting Rod	Music Wire, 0.015" x 4" (10.16 cm)
11	Displacer Pushrod	Music Wire 0.015" x 6" (15.24 cm)
12	Displacer Pushrod Flex Joint	Duct Tape, 1/8" (3.18 mm) x 1/2" (12.7 mm), 2 pieces
13	Drive Crankshaft Pin	Music Wire, 0.0625 x 1/2" (12.7 mm)
14	Drive Crankshaft Bushing	AWG-14 (0.066") Heavy Wall Extruded Teflon Tube, 3/8" (9.53 mm)
15	Drive Crankshaft Collar	Shaft Collar, 1/16" (1.59 mm)
16	Drive Crankshaft Plate	Aluminum Sheet, 0.062" 1/2" (12.7 mm) x 1" (25.4 mm)
17	Gland	Teflon Tube, #24 (0.022" ID) x 7/16" (11.11 mm)
18	Drive Cylinder	PVC Pipe, 1" (25.4 mm) ID x 5/8" (15.88 mm)
19	Drive Diaphragm	Latex glove fingertip
20	Drive Pushrod	Music Wire, 0.015" x 4" (10.16 cm)
21	Axle	Music Wire, 0.0625"x 1 1/4" (31.75 mm)
22	Axle Bushings (2 pieces)	AWG-14 (0.066") Heavy Wall Extruded Teflon Tube, 3/16" (4.76 mm)

*Flywheel thickness can vary slightly from the prescribed thickness of 0.06", depending upon the material available to the builder. Acrylic sheet commonly sold in the US market for window glazing is slightly thicker (0.093") and is used for the flywheels of some of the engines in this book. Thin polystyrene sheet material used in covering framed pictures is only 0.05" thick and will suffice, but can be difficult to work with.

** It is also possible to make the sidewalls from slightly thinner 3/16" (4.76 mm) acrylic sheet.

Part #8 is a round sidewall that is formed by heating a straight piece of acrylic sheet and bending it around a form.

Drawings and Dimensions

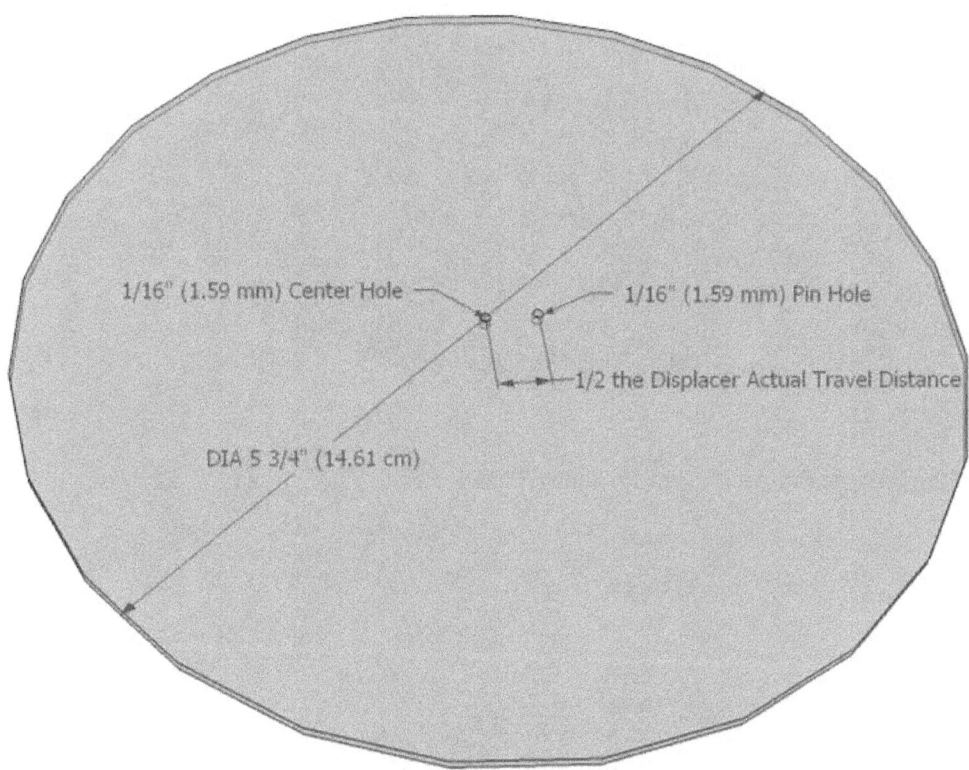

1/16" (1.59 mm) Center Hole

1/16" (1.59 mm) Pin Hole

1/2 the Displacer Actual Travel Distance

DIA 5 3/4" (14.61 cm)

Part # 01 Flywheel - Clear Acrylic Sheet, round, 0.06" x 5 3/4" (14.61 cm) diameter

Part # 02 Flywheel Collar - 1/16" (1.59 mm) Shaft Collar
Part # 16 Drive Crankshaft Collar - 1/16" (1.59 mm) Shaft Collar

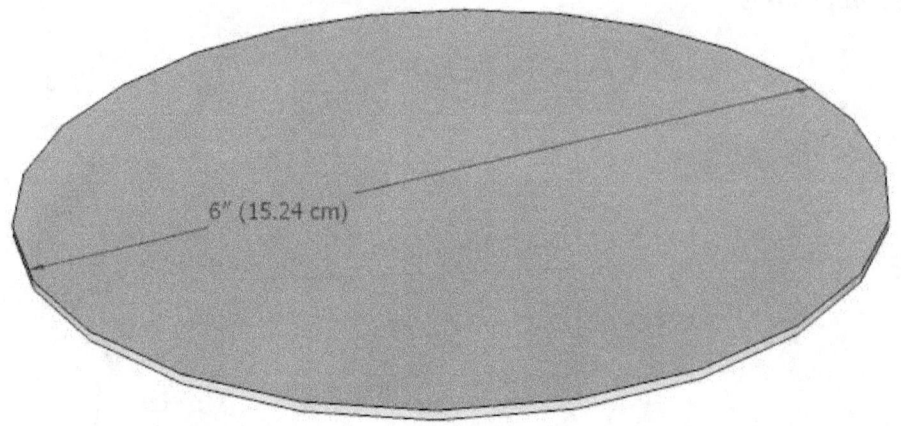

Part # 03 Pressure Chamber Bottom Plate - Aluminum Sheet, 0.062" x 6" (15.24 cm) x 6" (15.24 cm)

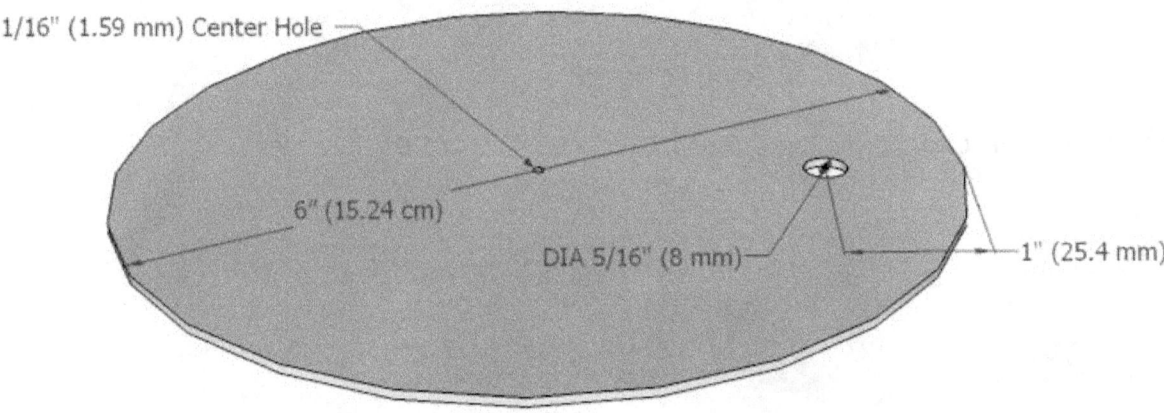

Part # 04 Pressure Chamber Top Plate - Aluminum Sheet, 0.062" x 6" (15.24 cm) x 6" (15.24 cm)

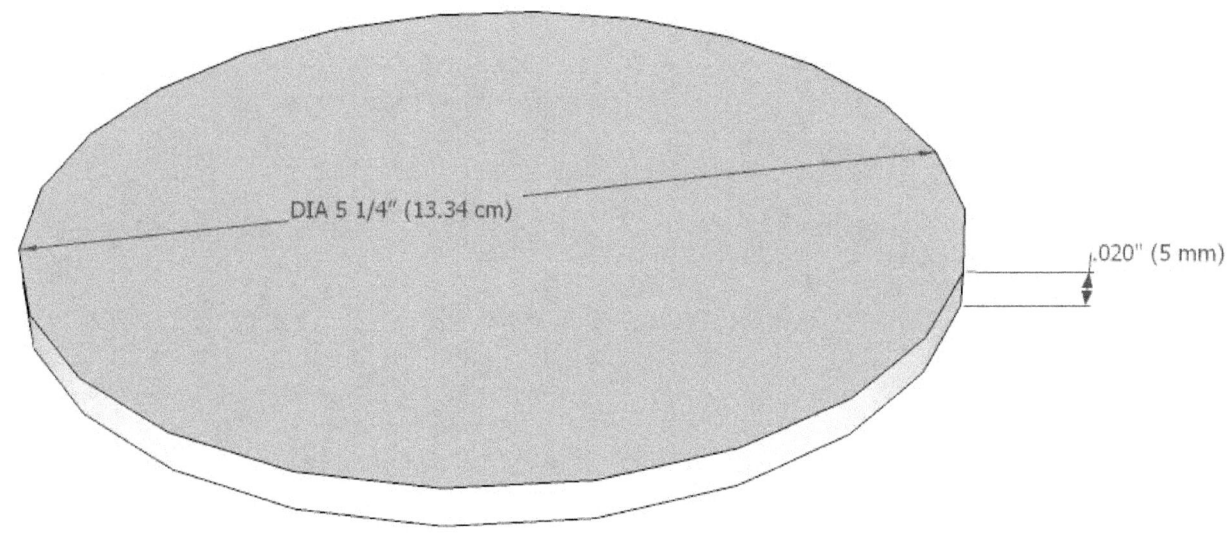

DIA 5 1/4" (13.34 cm)

.020" (5 mm)

Part # 05 Displacer - Styrofoam, 0.20" x 5 1/4" (13.34 cm)

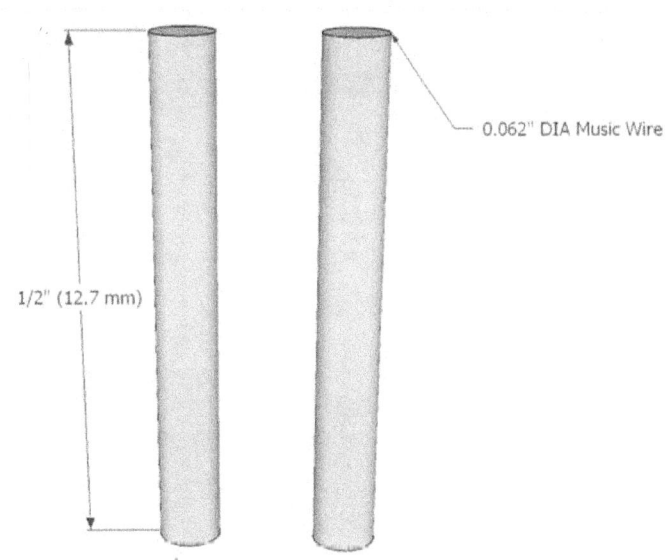

1/2" (12.7 mm)

0.062" DIA Music Wire

Part # 06 Displacer Crankshaft Pin - Music Wire, 0.0625 x 1/2" (12.7 mm)
Part # 14 Drive Crankshaft Pin - Music Wire, 0.0625 x 1/2" (12.7 mm)

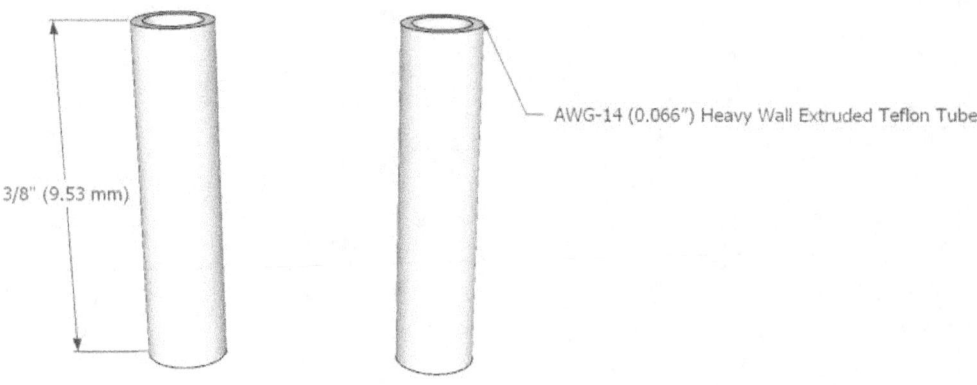

Part # 07 Displacer Crankshaft Bushing - AWG-14 (0.066") Heavy Wall Extruded Teflon Tube, 3/8" (9.53 mm)

Part # 15 Drive Crankshaft Bushing - AWG-14 (0.066") Heavy Wall Extruded Teflon Tube, 3/8" (9.53 mm)

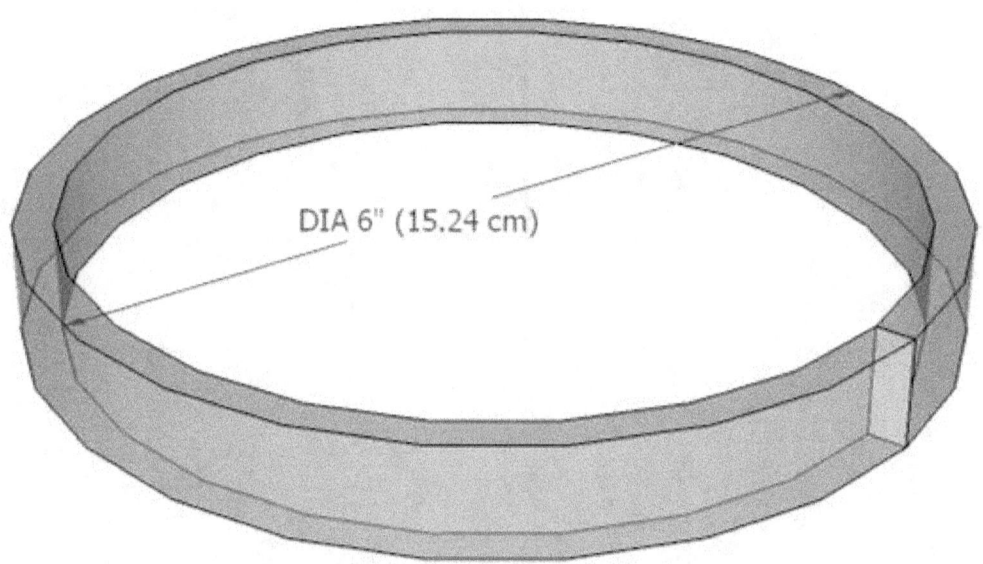

Part # 08 Pressure Chamber Sidewall - Formed Clear Acrylic, 1/4" (6.35 mm) x 11/16" (17.46 mm) x approx. 20" (50.8 cm)

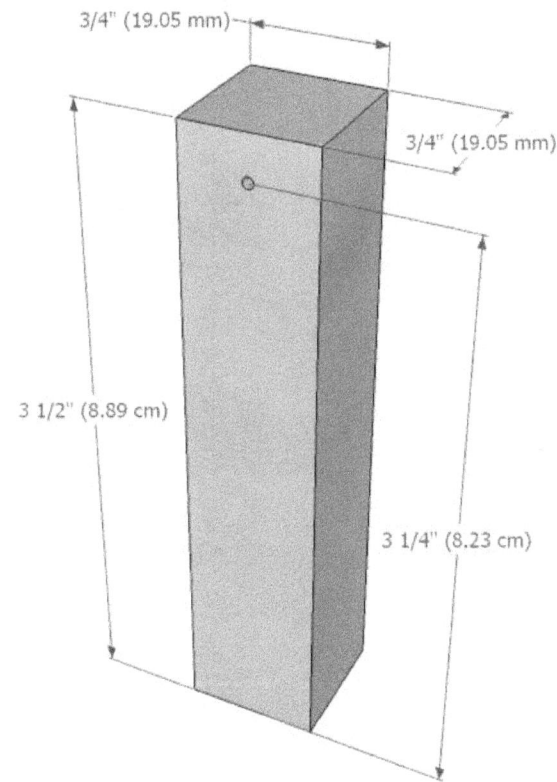

Part # 09 Pedestal - Wood, 3/4" (19.05 mm) x 3/4" (19.05 mm) x 3 1/2" (8.89 cm)

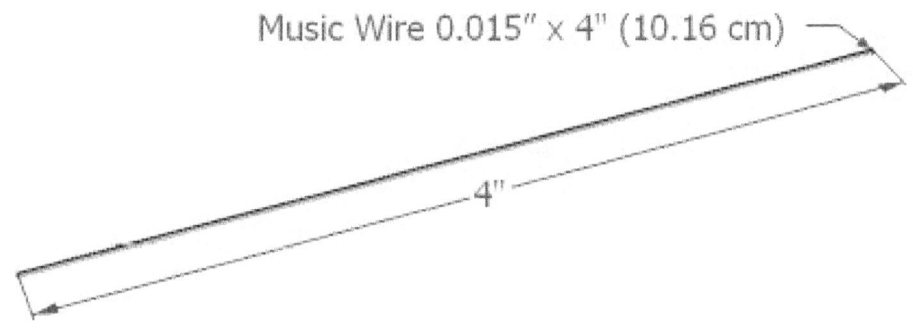

Part # 10 Displacer Connecting Rod - Music Wire, 0.015" x 4" (10.16 cm) and,
Part # 20 Drive Pushrod Music Wire, 0.015" x 4" (10.16 cm)
(2 Pieces)

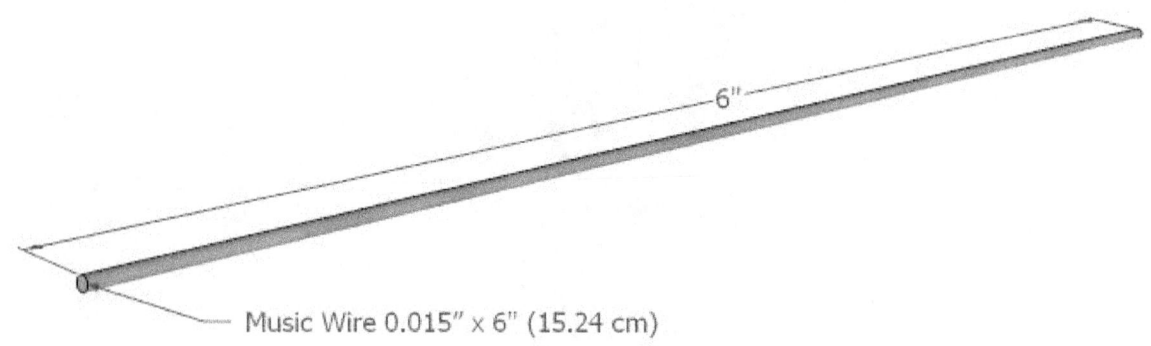

Music Wire 0.015" x 6" (15.24 cm)

Part # 11 Displacer Pushrod - Music Wire 0.015" x 6" (15.24 cm)

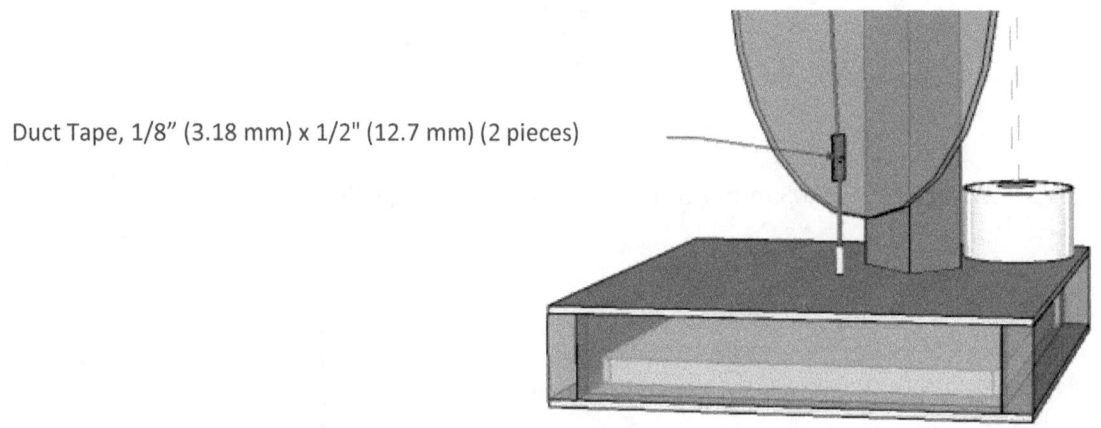

Duct Tape, 1/8" (3.18 mm) x 1/2" (12.7 mm) (2 pieces)

Part # 12 Displacer Pushrod Flex Joint - Duct Tape, 1/8" (3.18 mm) x 1/2" (12.7 mm) (2 pieces)

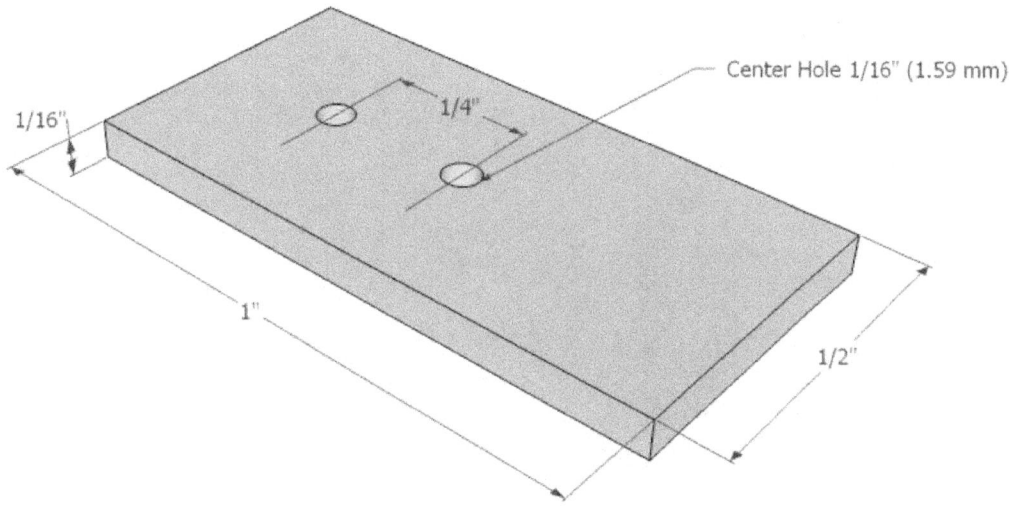

Part # 16 Drive Crankshaft Plate - Aluminum Sheet, 0.062" (1/16", 1.59 mm), 1/2" (12.7 mm) x 1" (25.4 mm)

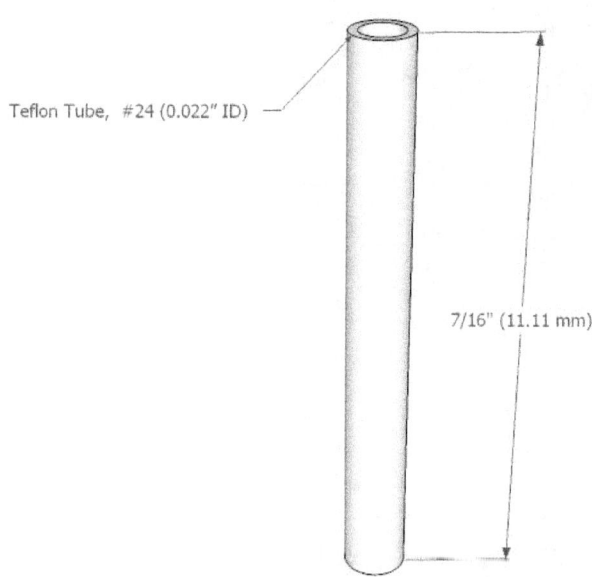

Part # 17 Gland - Teflon Tube, #24 (0.022" ID) x 7/16" (11.11 mm)

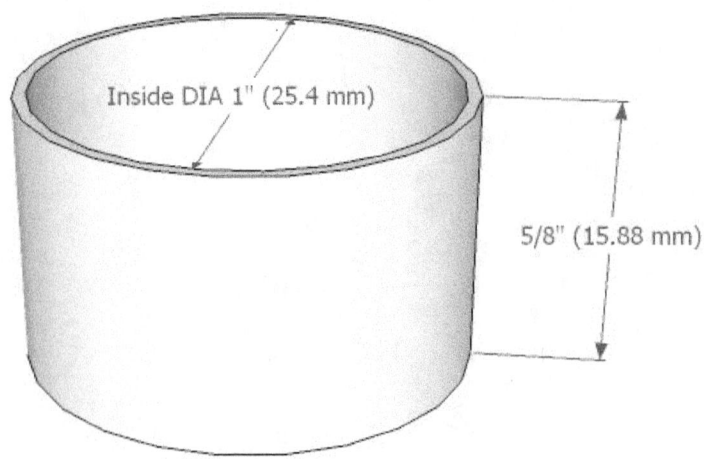

Inside DIA 1" (25.4 mm)

5/8" (15.88 mm)

Part # 18 Drive Cylinder - PVC Pipe, 1" (25.4 mm) ID x 5/8" (15.88 mm)

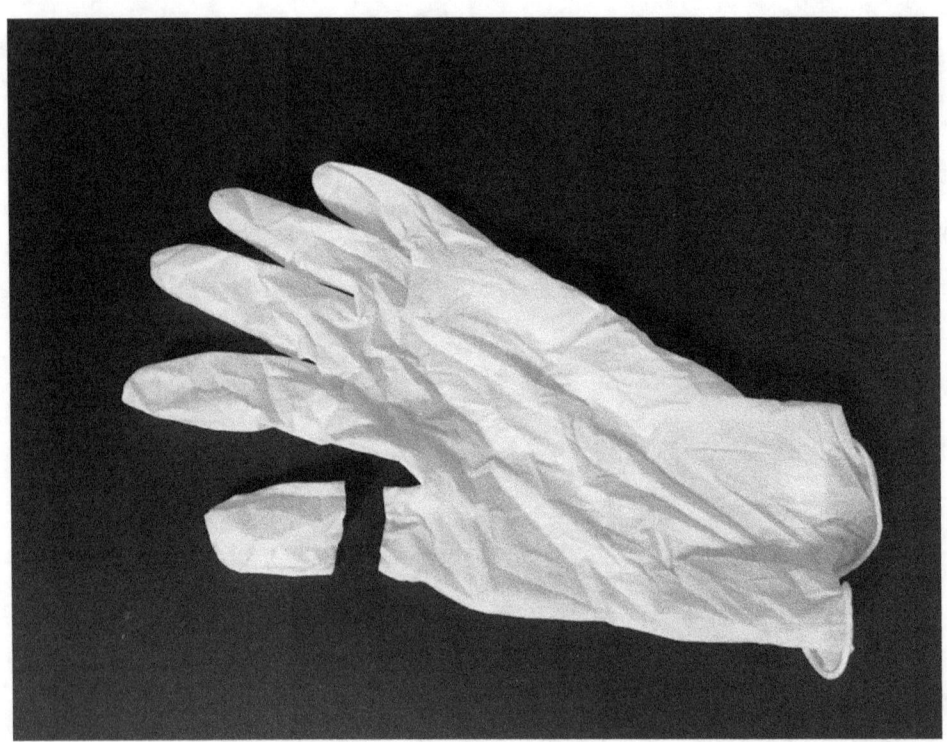

Part # 19 Drive Diaphragm - Latex glove fingertip

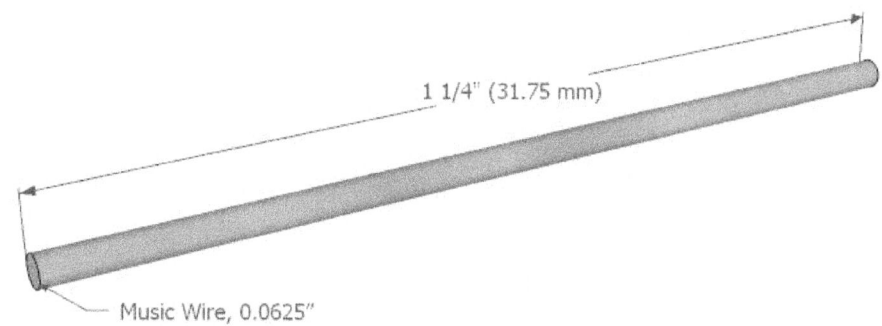

1 1/4" (31.75 mm)

Music Wire, 0.0625"

Part # 21 Axle - Music Wire, 0.0625"x 1 1/4" (31.75 mm)

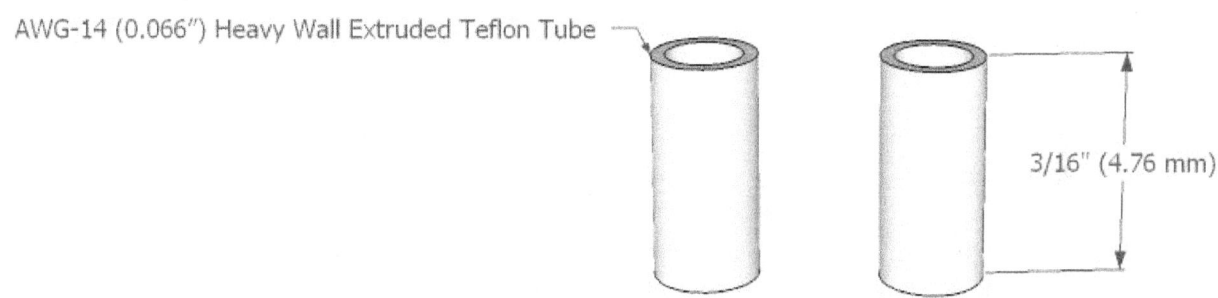

AWG-14 (0.066") Heavy Wall Extruded Teflon Tube

3/16" (4.76 mm)

Part # 22 Axle Bushings - AWG-14 (0.066") Heavy Wall Extruded Teflon Tube, 3/16" (4.76 mm) (2 pieces)

Assembly Instructions for Large Round Engine

Make the Pressure Chamber Top and Bottom Plates

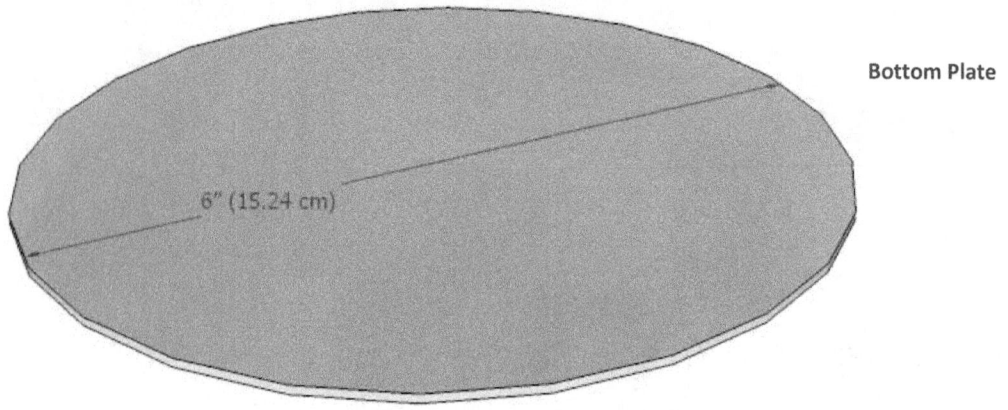

Bottom Plate

6" (15.24 cm)

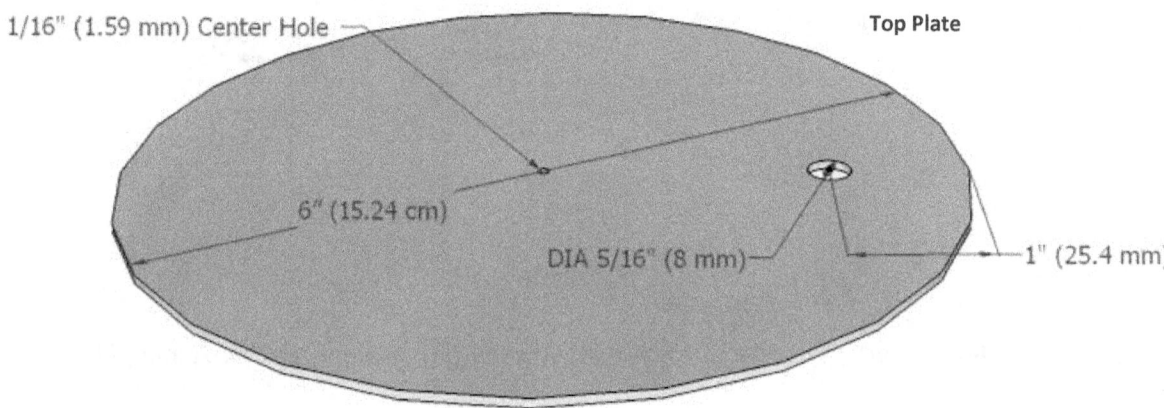

1/16" (1.59 mm) Center Hole

Top Plate

6" (15.24 cm)

DIA 5/16" (8 mm)

1" (25.4 mm)

Cut two aluminum plates to form the top and bottom plates of the pressure chamber. Plate thickness is 0.0625" (1.59 mm). The plates are round, measuring 6" (15.24 cm) in diameter

Wash the metal parts with soap and water, rinse well and allow them to dry.

Paint all surfaces of the top and bottom plates (if desired). Use spray enamel that is suitable for painting metal.

Mark the center point of the top plate when laying out the pattern on the material. Use a center punch to mark the spot. Drill a hole that is the same diameter (or slightly larger) than the outside diameter of the Teflon tube that will be used for the displacer gland. The size will vary depending upon the material chosen.

Mark the position for the drive cylinder hole. The hole is located 1" (25.4 mm) from the side of the top plate. The size of the hole is 5/16" (7.94 mm). Mark the hole position with a center punch and then carefully drill the hole.

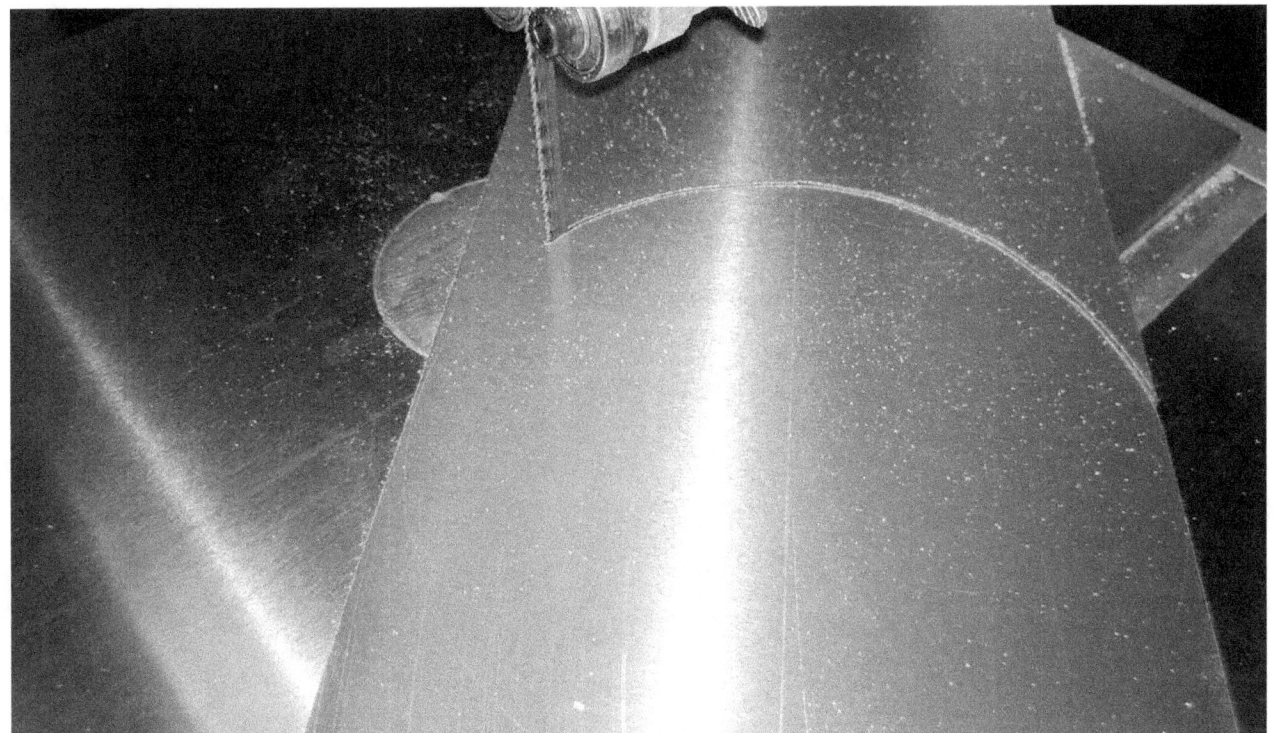

The top plates for these engines were cut with a band saw.

Make the Pressure Chamber Side Wall

The pressure chamber side wall is made by heating and bending a piece of clear acrylic. This process is sometimes called thermoforming. This process will require these three things:

- A piece of clear acrylic sheet measuring 1/4" (6.35 mm) x 11/16" (17.46 mm) x approximately 20" (50.8 cm) long.
- A heat gun to warm the acrylic.
- A jig or form to shape the acrylic and hold it in the proper shape until it cools.

Cut the Material to Length

The sidewall can be made from a single piece of acrylic, or it can be made in sections and pieced together. The examples shown here were made in sections and pieced together because the material available was not long enough to form the side from a single piece.

The length of the sidewall will be the same as the circumference of the top and bottom plates. The circumference of the plates can be calculated by multiplying the diameter of the circle by 3.1416.

6" x 3.1416 = 18.85", or 18 7/8"

15.24 cm x 3.1416 = 47.88 cm

The acrylic sheet material should be about 1/4" (6.35 mm) thick. Thinner material may be substituted. The thickness of the material used to make the sidewall will determine the size of the form that needs to be made for shaping the material during thermoforming.

Build a Form

The form is a round disk that is the same size as the inside diameter of the finished sidewall. This dimension is calculated by subtracting the thickness of the sidewall material (times 2) from the diameter of the pressure chamber. So for 1/4" (6.35 mm) material the inside diameter of the sidewall will be 5 1/2" (13.97 cm).

This picture shows a form built for bending the sidewall to make the 6" (15.24 cm) engine. The holes make it possible to use spring clamps to hold the material against the form.

Shape the Acrylic Sidewall with Heat

A heat gun is used to warm the acrylic sheet until it becomes soft and pliable. Use extreme caution and work in a well ventilated area. Acrylic will generate poisonous fumes if overheated, and it is flammable. Use just enough heat to soften and bend the acrylic, and no more. The heat gun is hot enough to cause severe burns or fire, so use caution with this tool.

Use a clamp to hold one end of the acrylic strip against the form. Apply heat to the first several inches of material. Apply occasional light pressure against the material to test it. It will eventually become soft with a consistency similar to clay.

Continue to apply heat ahead of the area to be bent as the acrylic is turned around the shape of the form.

This picture shows a strip of acrylic sheet as it is clamped in place in preparation for bending. The protective covering will have to be removed from the acrylic sheet before heat is applied.

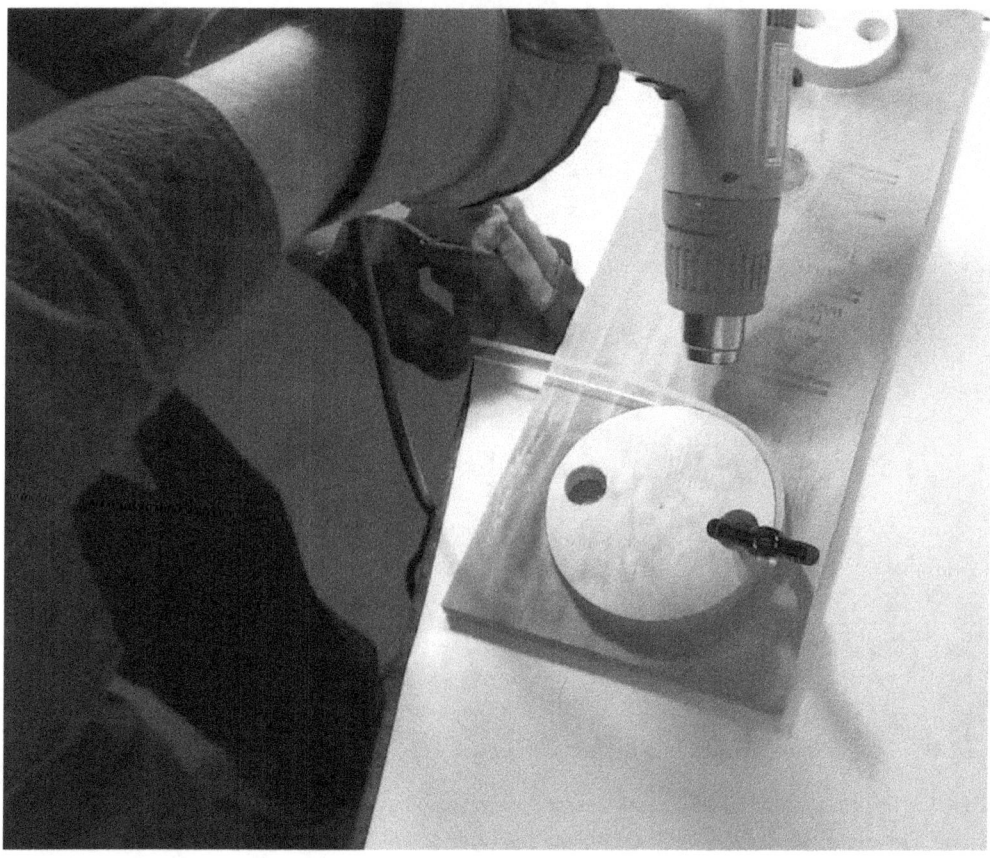

Apply heat to soften the material and work around the form.

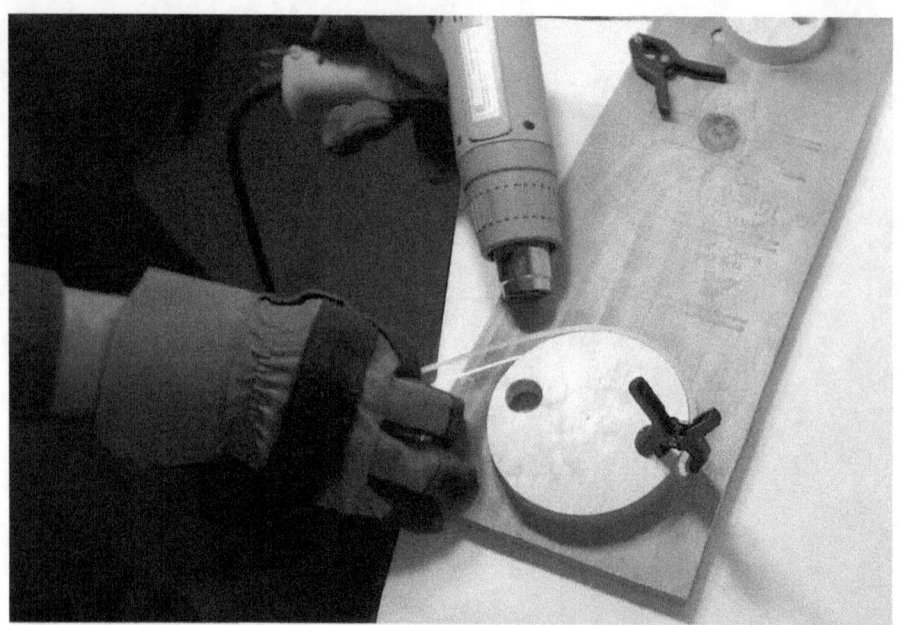

Continue heating the material as it is bent to the shape of the form.

When finished, allow the material to remain against the form until the acrylic has cooled and become rigid again. Since this piece was not long enough to complete the circle, it will be joined with another piece to make a complete sidewall.

Attach the Sidewall to the Pressure Chamber Bottom Plate

Carefully check the fit of all the parts before assembly and use enough adhesive to ensure a good seal. If multiple pieces were made to form the sidewall, they can be joined together during this process of attaching them to the bottom plate. Use clear five-minute epoxy to attach the acrylic sidewall to the pressure chamber bottom plate.

The sidewall must be attached so that the pressure chamber is air-tight. A small bead of glue oozing out of the joint indicates that the joint is being sealed well. Too much oozing glue, however, may interfere with the movement of the displacer inside the pressure chamber.

The top plate will _not_ be glued on at this time. The top plate may be used to help hold the sidewall in place as the glue hardens.

Note: The glue will leave a permanent mark anywhere that it comes in contact with the acrylic. Be very careful not to create too many fingerprints on the acrylic during the gluing process, as these may be permanent.

Note: Silicone adhesive is another option for gluing the sidewalls to the pressure chamber. Silicone adhesives tolerate heat better than most clear epoxy glues, but they do not hold as well. Silicone is especially useful if there is a need to disassemble the engine for any reason.

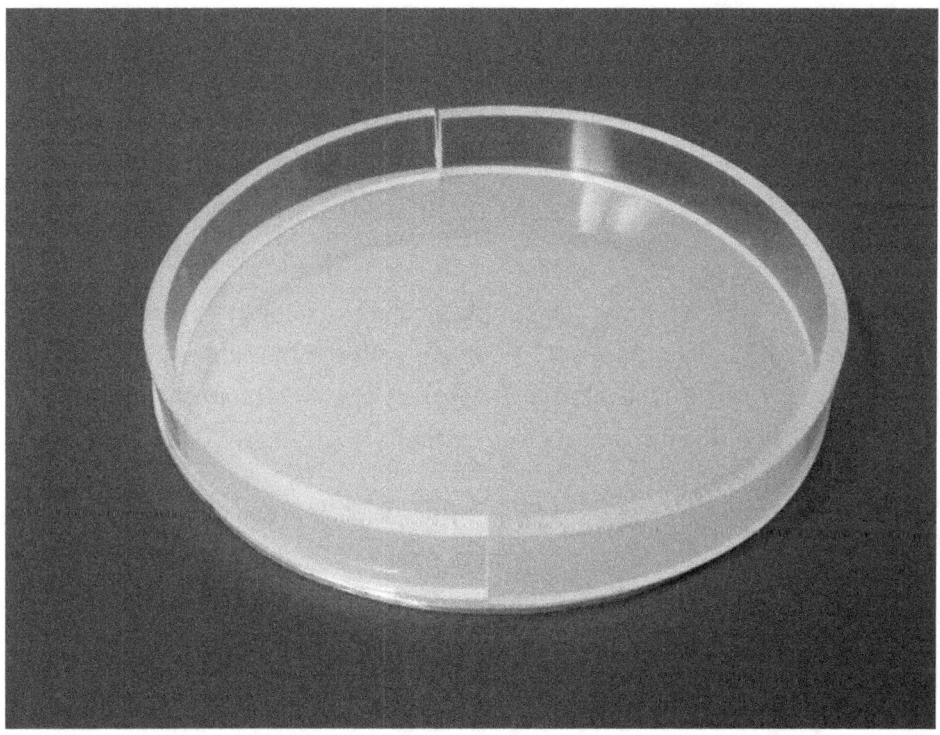

The pressure chamber bottom plate and sidewalls are ready for assembly. Dry fit the parts before gluing and make any adjustment necessary to create a good fit.

Cut the Foam Displacer Panel

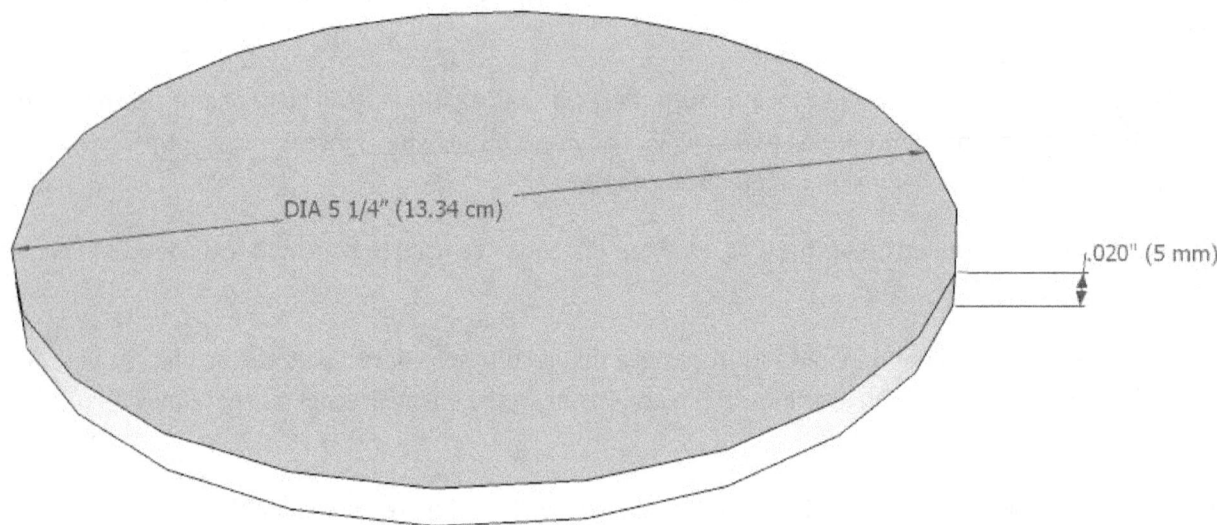

The displacer is made from Styrofoam that is cut to a thickness of 0.20" (13/64", 5.08 mm). For the best results, cut the foam with a hot wire foam cutter similar to the one described elsewhere in this book.

A hot wire foam cutter uses Nichrome wire that is heated by a small electrical current to cut the foam. The framework of the hot wire foam cutter holds tension on the wire to keep it straight, and a small electrical transformer provides the current to heat the wire. Foam cutters can be purchased from craft suppliers, or one can be built using simple parts. Plans for a homemade hot wire foam cutter are provided later in this book.

The dimensions of the displacer provide clearance between the displacer and the sidewall of the pressure chamber. That clearance needs to be 1/16" (1.59 mm) to 1/8" (3.18 mm) between the side of the displacer panel and the pressure chamber sidewall. Measure the internal dimensions of the pressure chamber and make any necessary adjustments to the size of the displacer before cutting.

Use a compass to create the circle on the foam sheet and mark the center of the circle. The center point will be used to locate the displacer pushrod.

Attach the Displacer Pushrod
Cut a piece of 0.015" music wire to a length of approximately 6" (15.24 cm). Small diameter music wire can be cut with ordinary wire cutters.

Make a 90° bend 1/2" from one end of the music wire.

A simple jig is used to hold the wire at the correct angle while attaching it to the foam displacer. Make the jig by drilling a small hole through a piece of flat wood. Drilling the hole in the jig will require a small diameter drill bit and a drill press. Choose a bit size that will provide a snug fit for the music wire.

If small drill bits under 1/16" (1.59 mm) are not available, use a 1/16" (1.59 mm) drill bit and insert a piece of Teflon tube into the hole to make a better fit for the music wire.

Pierce the foam at the center point with the long straight end of the displacer pushrod and press it in until the 90⁰ bend is pressed up against the surface of the foam. Place the protruding music wire into the hole in the jig and pull it though the jig until the foam is setting flush against the surface of the jig. The jig should now be holding the pushrod perpendicular to the foam displacer.

Press down on the wire at the center of the displacer to make a small depression in the foam no more than 1/32" (0.79 mm) deep. Fill the depression with high temperature epoxy to attach the pushrod to the displacer. Leave the displacer and pushrod in the jig until the epoxy has cured.

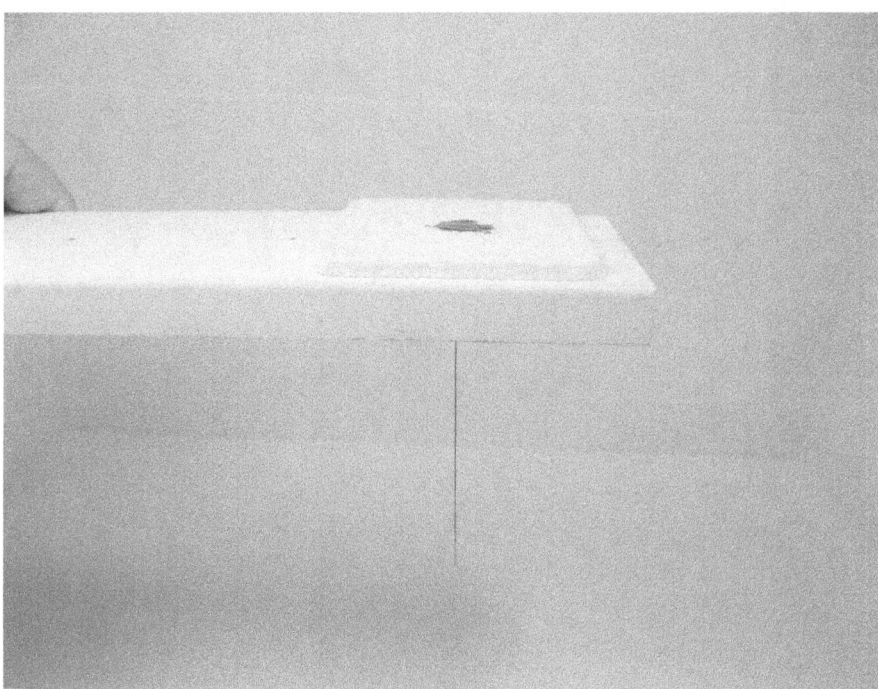

The displacer pushrod is held at the correct angle by the wooden jig. A hole is drilled through the board to hold the shaft at the correct angle. High temperature epoxy is used to attach the pushrod to the displacer panel. This picture shows a square displacer. The process is the same for the round displacer.

Prepare the Drive Cylinder Tube

As noted earlier, the tubing used to make the drive cylinder is approximately 1" (25.4 mm) in diameter. Cut a piece of this tubing to a length of 5/8" (15.88 mm). Cut the pipe carefully to maintain a 90⁰ angle on each end of the short drive cylinder.

Use fine sand paper to remove the glossy factory surface from the PVC pipe. This will also remove any labels and markings and improve the look of the finished part. Sanding the outer surface of the drive cylinder will help hold the latex drive diaphragm in place.

Place the drive cylinder over the drive cylinder hole on the pressure chamber top plate and glue it in place with epoxy. The drive cylinder is centered between the pedestal position and the edge of the pressure chamber top plate. Apply the glue on the inside edge of the drive cylinder tube. Applying the glue on the inside edge will hide the glue joint from view when the engine is fully assembled.

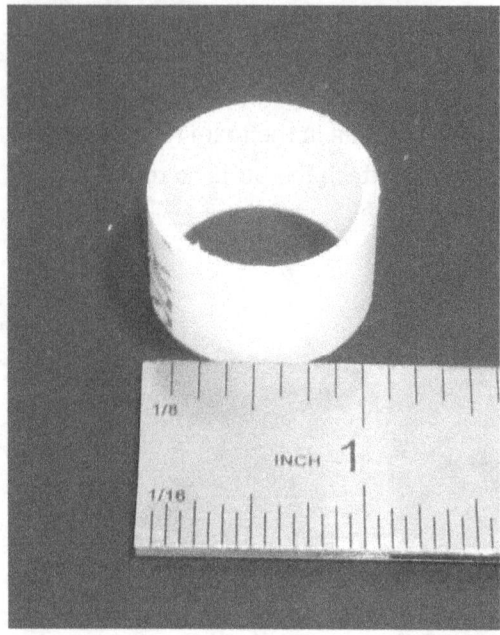

**The drive cylinder is made from a short piece of 1"
(25.4 mm) diameter PVC pipe.**

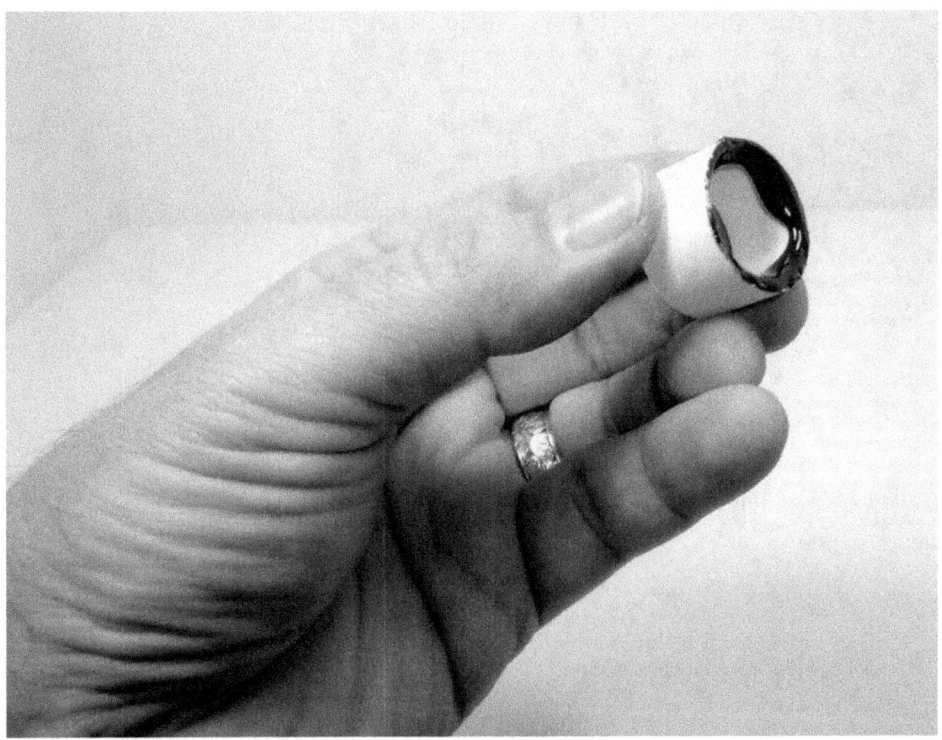

The drive cylinder is attached with epoxy. The glue is spread on the inside of the cylinder to create a nice looking airtight joint.

The drive cylinder is glued in place on the pressure chamber top. The glue is on the inside of the cylinder to provide an airtight seal that will be hidden from view after final assembly.

This picture shows the position of the drive cylinder on a finished engine.

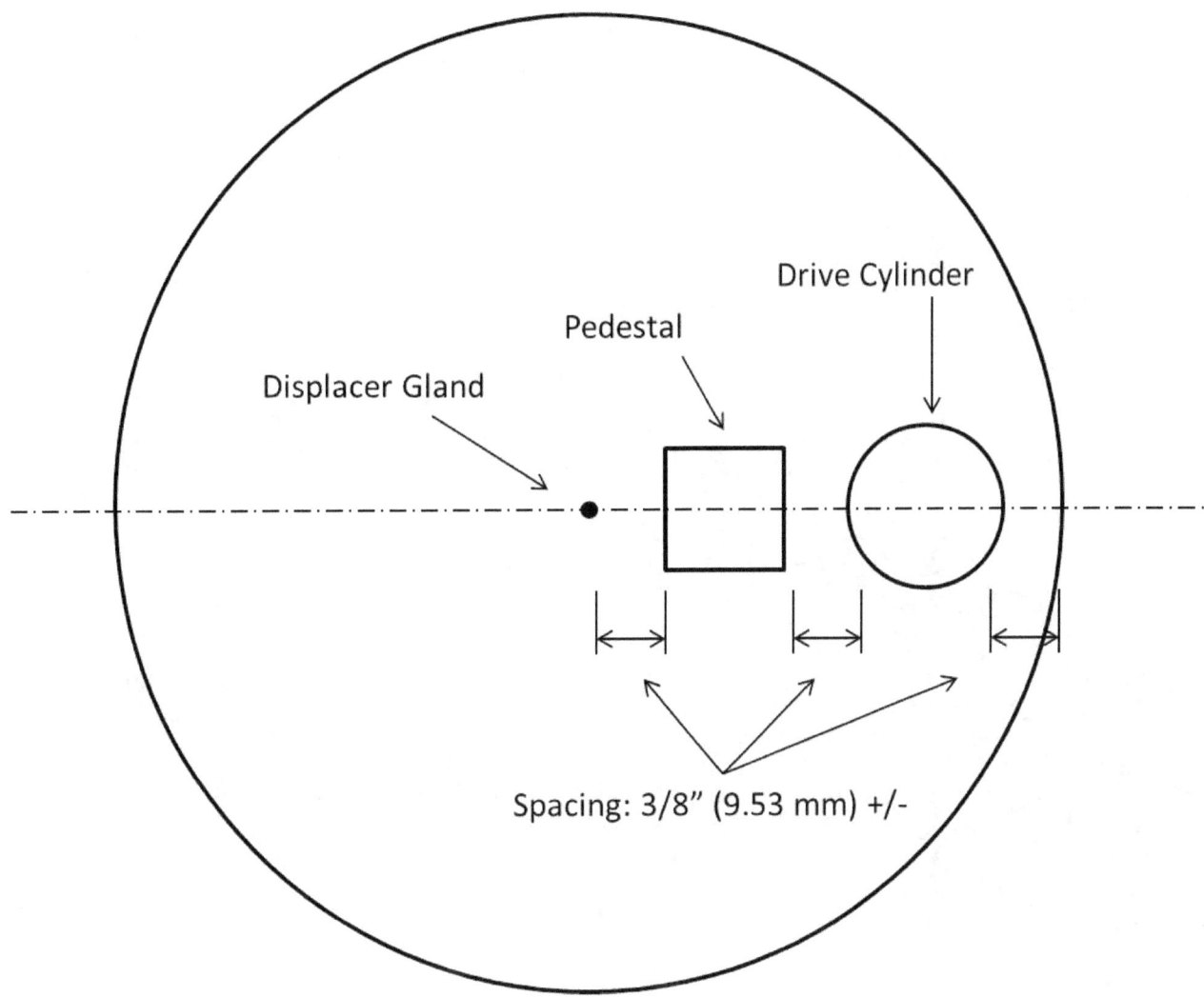

Drive Cylinder

Pedestal

Displacer Gland

Spacing: 3/8" (9.53 mm) +/-

This diagram shows the layout of the parts on the pressure chamber top.

Attach the Displacer Gland

In the world of machinery, a "gland" is a sleeve within a stuffing box, fitted over a shaft in such a way as to prevent leakage of fluid while allowing a shaft or stem to move. Additionally, the gland of a Stirling engine allows the shaft to move with little or no friction. The gland is nearly air-tight. A tiny (very tiny!) pressure leak is desirable and helps the engine adjust for pressure changes as it warms up.

The gland is made of #24 (0.022" ID) Teflon tube. This provides a friction free fit for the 0.015" displacer pushrod.

Cut a piece of this Teflon tube to a length of 7/16" (11.11 mm).

Cutting Teflon tube requires a very sharp knife or razor blade, and a piece of 0.015" music wire. Insert the music wire into the Teflon tube before cutting. Place the tubing and wire on a hard flat surface. Press the blade of the knife against the tubing and roll the tubing on the flat surface until the knife has cut all the way around the tube. The wire in the middle of the tube will stop the knife from cutting all the way through the tube unless the tube is rolled. This will prevent the end of the tube from being crushed or deformed during the cutting process.

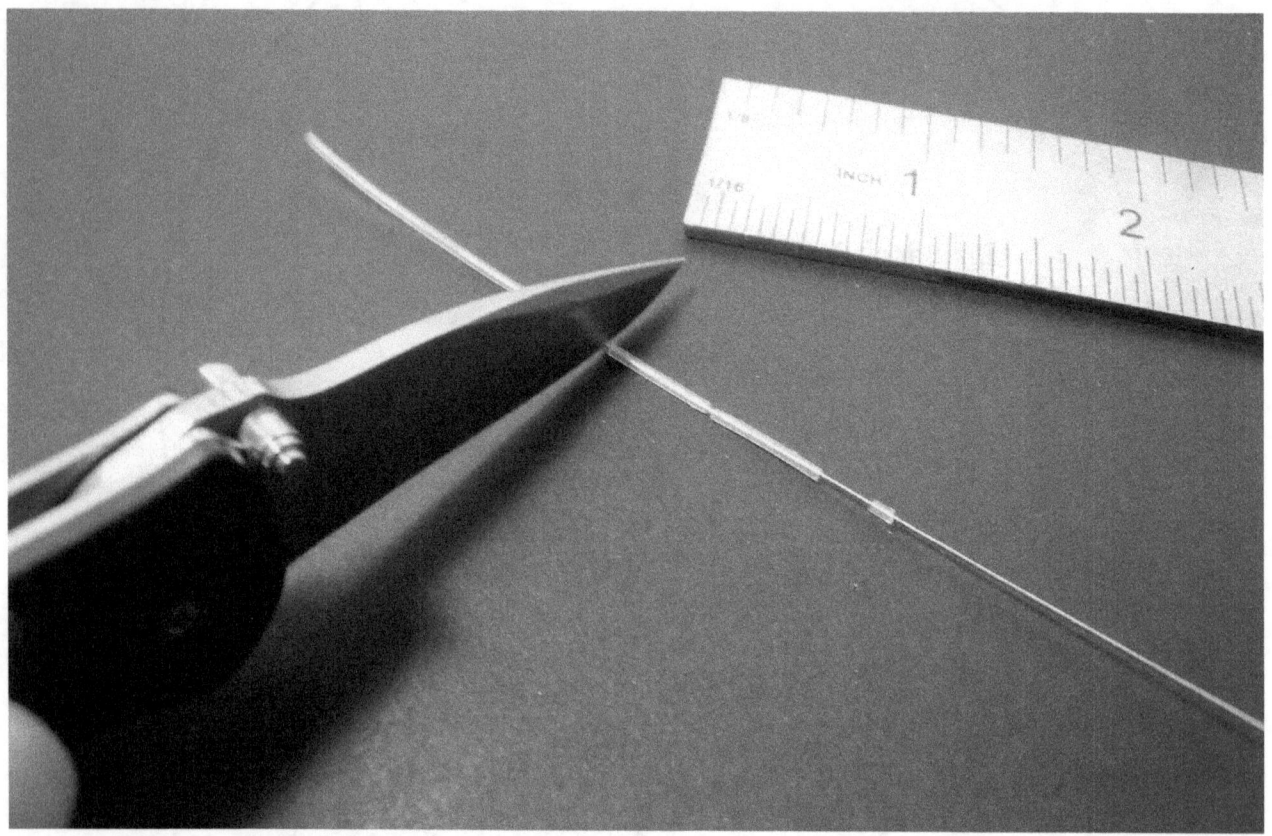

Insert a piece of music wire inside the Teflon tube before cutting. Roll the tube on a hard surface to cut around the wire. Cutting in this manner prevents the tube from deforming.

To attach the displacer gland:

1. Locate the displacer/pushrod assembled previously, the pressure chamber top plate, and the Teflon gland.
2. Slide the displacer pushrod through the top plate of the pressure chamber.
3. Set the displacer/top plate on a flat level surface with the pushrod pointing upward.
4. Slide the 7/16" (11.11 mm) gland tube onto the displacer pushrod.
5. Test for a friction free fit by verifying that the short piece of Teflon tube can fall under its own weight when dropped.
6. Let the Teflon gland drop inside the hole on the pressure chamber top plate until the end of the Teflon tube is flush with the inside surface of the top plate.
7. Apply a small bead of epoxy glue around the base of the gland tube, sealing it to the pressure chamber top plate. Take care that the glue does not touch the displacer pushrod. Allow the glue to cure.

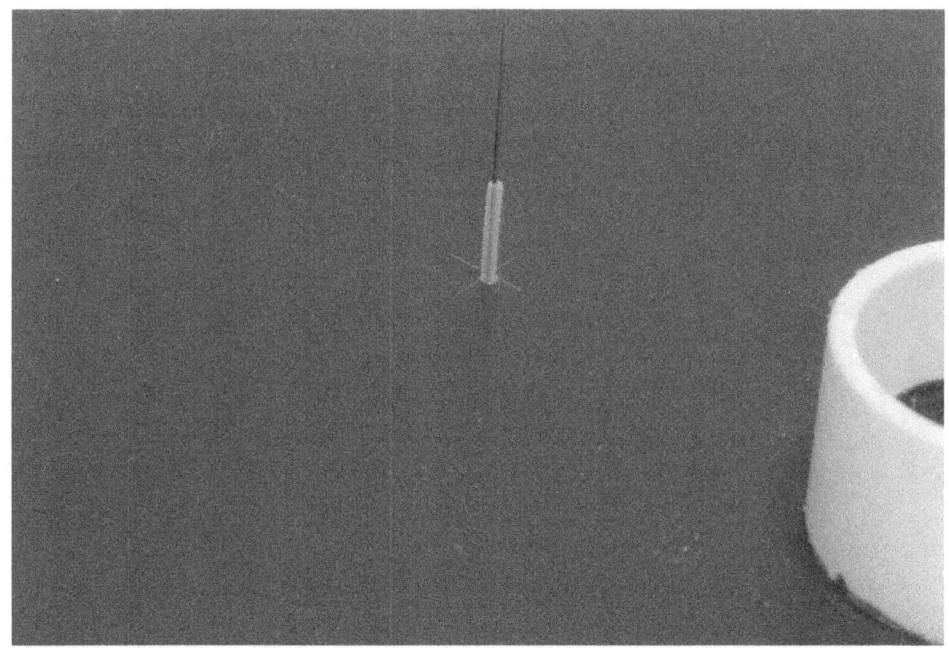

The pressure chamber top plate is placed over the displacer with the pushrod in place. The Teflon tube gland is shown here before glue is applied.

High temperature epoxy is applied around the base of the displacer gland.

Create the Flywheel

The flywheel is the largest and most visible moving part on this engine. If this part is built with care it will give the engine a very nice finished look. The goal is to make a flywheel that is perfectly round with a connection point for the axle that is squared to the flywheel and in the exact center. If the end result is less than perfect, the engine will still operate, and may even operate quite well. It just looks better if it is round, squared to the shaft, and centered.

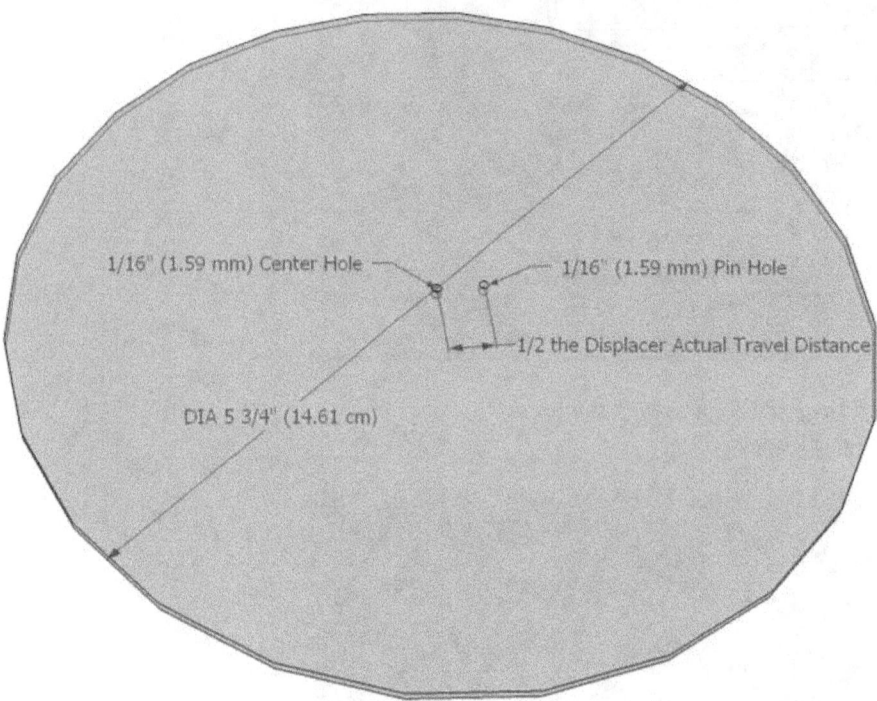

The flywheel for the large engine is 5 3/4" (14.61 cm) in diameter.

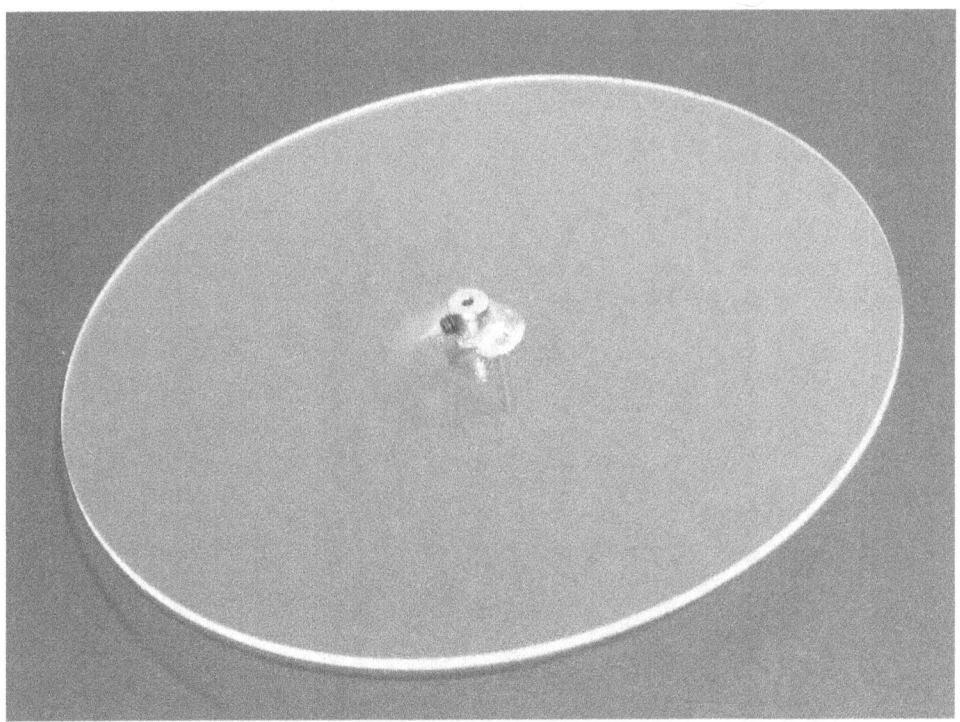

The flywheel pictured here was cut on a band saw using a simple homemade circle cutting jig. A shaft collar is attached at the center of the flywheel, and the displacer crankshaft pin is attached on the opposite side, slightly off center.

Make a Simple Circle Cutting Jig

The process for creating a circle cutting jig and cutting flywheel is shown in detail in the instructions for the 4" Square engine in a previous chapter of this book. The process for making this flywheel is the same, with the exception that the diameter of this flywheel is larger. Please refer to the previous chapter for detailed instructions for how to make the flywheel.

Using a band saw and a circle cutting jig is the recommended method for making the flywheel. If a band saw is not available, the flywheel can be cut with hand tools.

Assemble the Pressure Chamber with the Displacer

Place the displacer pushrod through the gland and place the top plate on the pressure chamber with the displacer inside. Check the fit of all the parts and ensure that the displacer can move up and down without being obstructed. Make any necessary adjustments to the displacer before the top is glued onto the pressure chamber.

Use silicone adhesive to attach the pressure chamber top plate of the round sidewall. Spread a thin layer of the adhesive on the top edge of the sidewall and smooth it with your finger. Use enough adhesive to create a seal, but no more. If too much adhesive is applied it may ooze into the inside of the pressure chamber and interfere with the motion of the displacer. Clamp the parts together with light pressure and set them aside until the silicone adhesive has cured.

Silicone adhesive is used to attach the top plate because it can be opened up later if repairs are necessary.

Drill the Hole for the Crankshaft Pin in the Flywheel

The crankshaft pin is mounted in a hole that is drilled near the center hole of the flywheel. The spacing between the center flywheel hole and the crankshaft pin determines how far the displacer will travel as the flywheel rotates. The spacing between these two holes will be exactly half the _actual travel distance_ of the displacer.

Measure the travel of the displacer pushrod. To do this, take a measurement from the top of the pressure chamber to the top of the displacer pushrod when the displacer is on the bottom of the pressure chamber. Now lift up on the pushrod until the displacer is at the top of the pressure chamber and measure it again. Subtract the smaller number from the larger number to calculate the _total available travel distance_.

Subtract 1/16" (1.59 mm) from the _total available travel distance_ to get the _actual travel distance_. The actual travel distance will be slightly shorter than the total available distance so that the displacer will not touch the top or bottom of the pressure chamber as the engine runs. Shortening the travel distance by 1/16" (1.59 mm) will provide 1/32" (0.79 mm) of clearance above and below the displacer and prevent it from coming into contact with the pressure chamber.

It is important that the displacer does not come into contact with the pressure chamber as the engine is running. If the displacer hits the pressure chamber this will increase friction or prevent the engine from rotating freely. Also, it is good to keep a small cushion of air between the displacer and the pressure chamber top and bottom plates. The cushion of air reduces drag from what some refer to as "pull-off friction."

Pull-off friction can be demonstrated by holding a flat piece of cardboard against the ceiling with a broom handle. If the broom handle is quickly removed the cardboard does not immediately drop. The air pressure on the bottom of the cardboard holds it up until air is able to get in between the cardboard and the ceiling and equalize both pressures. This same effect can happen inside the pressure chamber if the displacer comes to rest in contact with the top or bottom plate of the pressure chamber. For this reason the engine is designed so that the displacer clearance is between 1/32" (0.79 mm) and 1/16" (1.59 mm).

Once you have determined the _actual travel distance_ of the displacer, divide that distance in half. This number will be the distance between the center hole of the flywheel and the hole for the crankshaft pin. Drill the hole for the crankshaft pin with a 1/16" (1.59 mm) drill.

The same circle cutting jig that was used to make the flywheel on the band saw is also an excellent jig for measuring and drilling the hole for the displacer crankshaft pin. The illustration shows how a pair of dividers can be used to measure the distance from the center pin to find the location for the displacer crankshaft pin hole.

Set the dividers to the distance needed for the offset of the crankshaft pin. Center one divider point on the center pin and the other point on the middle of the drill bit. Drill the hole when the alignment is correct.

Attach the Shaft Collar to the Flywheel

The flywheel is attached to the axle by means of a small round shaft collar that contains a set screw. The shaft collar is glued to one side of the flywheel in the exact center. Great care needs to be taken to align the shaft collar so that the main axle is perpendicular to the surface of the flywheel.

An alignment jig similar to what was used to attach the displacer pushrod will help attach the shaft collar with good alignment. Use a drill press to drill a 1/16" (1.59 mm) hole though a flat board. This board will serve as the alignment jig. Place a short piece of 1/16" (1.59 mm) music wire through the hole in the board. The music wire needs to be long enough to pass all the way through the board, the flywheel, and the shaft collar.

Place the flywheel over the music wire and press it flat against the surface of the board. Place the shaft collar over the music wire and press it flat against the flywheel. If the dry fit looks good, lift the shaft collar and spread a small drop of 5 minute epoxy under the shaft collar and press it against the surface of the flywheel.

The music wire will have to be removed before the epoxy is completely cured or it may become permanently attached. When the epoxy begins to harden, pull the music wire down through the alignment jig from the back side. This should leave the shaft collar glued over the center hole of the flywheel in near perfect perpendicular alignment.

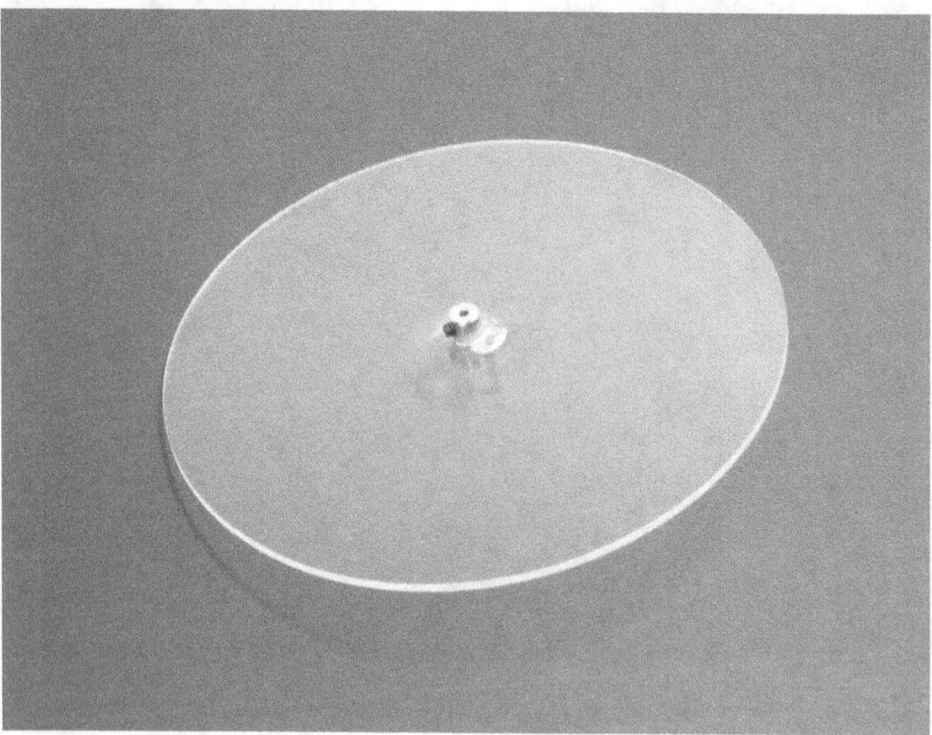

The shaft collar is attached over the center hole of the flywheel.

Make and Attach the Displacer Crankshaft Pin

There are two crankshaft pins. The displacer crankshaft pin attaches to a hole in the flywheel. The drive crankshaft pin is on the opposite end of the axle, over the drive diaphragm.

The displacer crankshaft pin is made from 0.0625" music wire, and is 1/2" (12.7 mm) long. The surface of the music wire should be smooth, with no gouges or tool marks, and it must be straight. Mark the wire where it is to be cut, clamp it in a vise, and then use the corner of a file to score the wire at the mark on two sides. Padding the vise and the pliers with paper will reduce the chances of scratching the music wire. Carefully bend the wire with pliers and it will break at the scored mark. Use a file or sandpaper to smooth the ends of the pin.

Glue the pin into the displacer pin hole on the flywheel. The location of the hole was calculated earlier. Use a small amount of epoxy to attach the pin to the flywheel. The pin must be perpendicular to the surface of the flywheel.

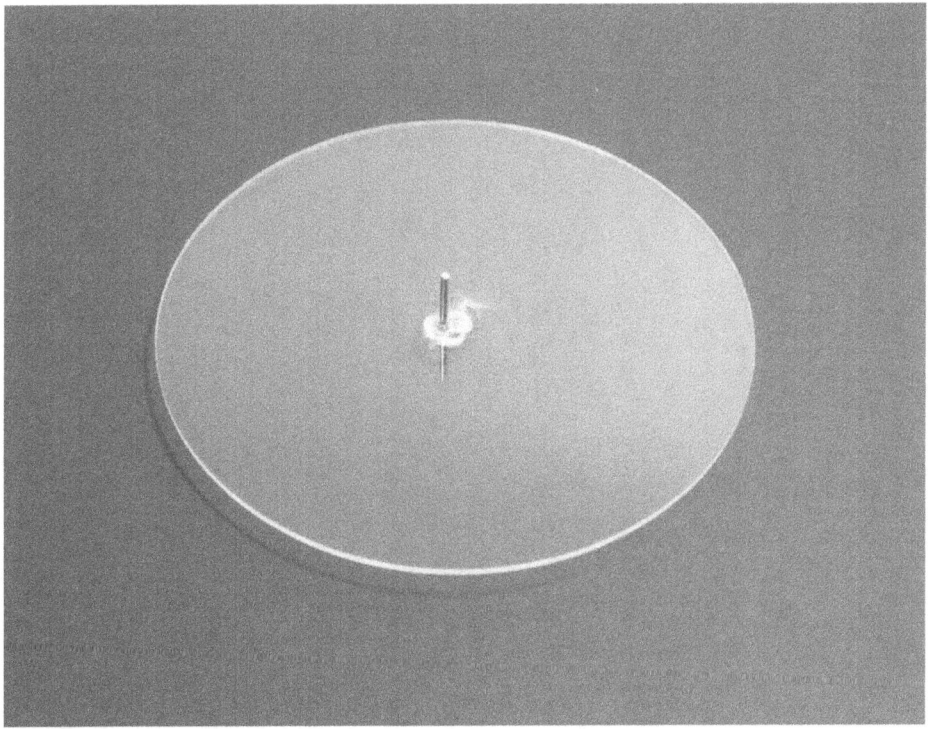

The crankshaft pin is attached to the flywheel on the opposite side as the shaft collar. The crankshaft pin is in the hole that is slightly off center.

Make the Main Axle

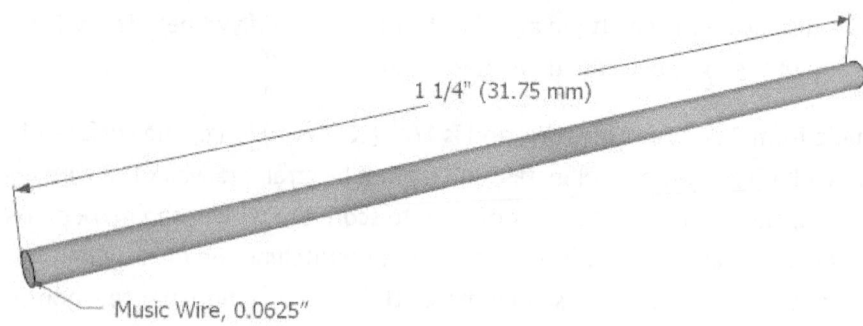

1 1/4" (31.75 mm)

Music Wire, 0.0625"

The axle is also made of 0.0625" music wire cut to a length of 1 1/4" (31.75 mm). Measure, score, and break the axle wire using the same technique that was used to cut the displacer crankshaft pin. Use a file or sandpaper to smooth the ends of the axle.

Make the Pedestal

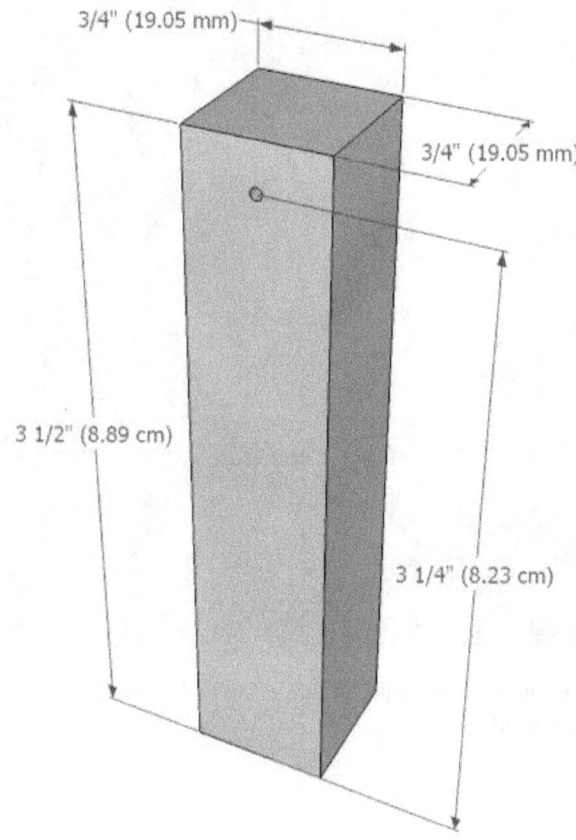

3/4" (19.05 mm)

3/4" (19.05 mm)

3 1/2" (8.89 cm)

3 1/4" (8.23 cm)

The pedestal is a small post that attaches to the top of the pressure chamber. It can be cut from a piece of wood. It measures 3/4" (19.05 mm) square by 3 1/2" (8.89 cm) long.

Drill a hole in the top of the pedestal for the axle. Position the hole 3 1/4" (8.29 cm) from the bottom of the pedestal (which is 1/4" (6.35 mm) below the top of the pedestal). The size of the hole must provide a snug fit for the Teflon tube used for the axle bushings. The tubing used in these illustrations fits snugly in a 7/64" (2.78 mm) hole. Measure to verify the hole size needed for the bushing material that will be used.

The pedestal may be finished with paint or varnish to improve the appearance of the finished engine.

Cut and Mount Teflon Tube Axle Bushings

The axle bushings are made from AWG-14 (0.066") Heavy Wall Extruded Teflon Tube. Two pieces are required. Each piece is 3/16" (4.76 mm) long.

Cut the tubing by first placing a piece of 0.0625" music wire inside the tube at the place to be cut. Place the tubing on a flat surface and roll the tubing as you cut it with a sharp knife or razor blade. Cut down to the music wire as the tubing is rolled on the flat surface. Cutting in this manner prevents the tubing from being deformed during the cutting process. Cutting the tubing without the shaft inside can cause the tubing to flatten or kink at the point of the cut, causing friction in the bushing.

Insert the small pieces of Teflon tube into each end of the hole in the pedestal. Leave a small amount of tubing (about 1/16" (1.59 mm) or less) protruding from the hole on both sides. No gluing will be required if the hole is the correct size.

The outside diameter of the axle is 0.0625", which allows it to turn freely inside the 0.066" ID Teflon tube. Insert a piece of axle material into the bushings to test the fit. Make any adjustments necessary to enable the axle to spin freely in the bushings.

The pedestal is shown here with the Teflon bushings ready to be installed.

Dry-Assemble and Measure for Locating the Pedestal

Attach the axle to the flywheel using the shaft collar, and insert this assembly into the bushings of the pedestal.

Position the pedestal on top the pressure chamber between the drive cylinder and the displacer pushrod. The axle must be centered over the displacer pushrod and the drive cylinder. The pedestal must be positioned so that both the pushrods can be attached to their respective pins at a 90° angle to the axle. Once the ideal position for the pedestal has been found, mark the position with a pencil. Set the pedestal aside for now. It will be attached after the drive crankshaft has been assembled.

Make the Drive Crankshaft

The drive crankshaft plate creates the offset for the crankshaft that attaches to the drive diaphragm. It holds the drive crankshaft pin parallel to the axle with an offset of 1/4" (6.35 mm). A hole is drilled to attach the crankshaft pin to one side. A shaft collar is glued to the opposite side and is the attachment point for the main axle.

Draw the shape of the crankshaft plate on a piece of aluminum stock and drill the hole before cutting the plate. The plate is made from 0.062" aluminum sheet, the same material recommended for the top and

214

bottom plates of the pressure chamber. The dimensions of the plate are 1/2" (12.7 mm) x 1" (25.4 mm). Drill a 1/16" (1.59 mm) hole at a position 1/4" (6.35 mm) from the center of the plate. Cut the small plate from the aluminum sheet after the hole has been drilled.

Cut the drive crankshaft pin from a piece of 0.062" music wire. The length of the pin is 1/2" (12.7 mm). Attach the pin to the hole in the plate with epoxy. The pin must be perpendicular to the surface of the plate. The pin is mounted on the front side of the plate.

Place a mark at the center on the back side of the plate, 1/4" (6.35 mm) away from the pin. This mark will be used to position the shaft collar to the back side of the plate. Use epoxy to attach a 1/16" (1.59 mm) shaft collar to the back side of the plate. Take care that the set screw of the shaft collar does not become fouled with epoxy.

The pictures show multiple crankshaft plates. Only one is required. Making several crankshaft plates with different offset measurements makes it possible to change the travel distance of the drive mechanism. This can be a useful adjustment for fine tuning the engine to operate in different temperature environments.

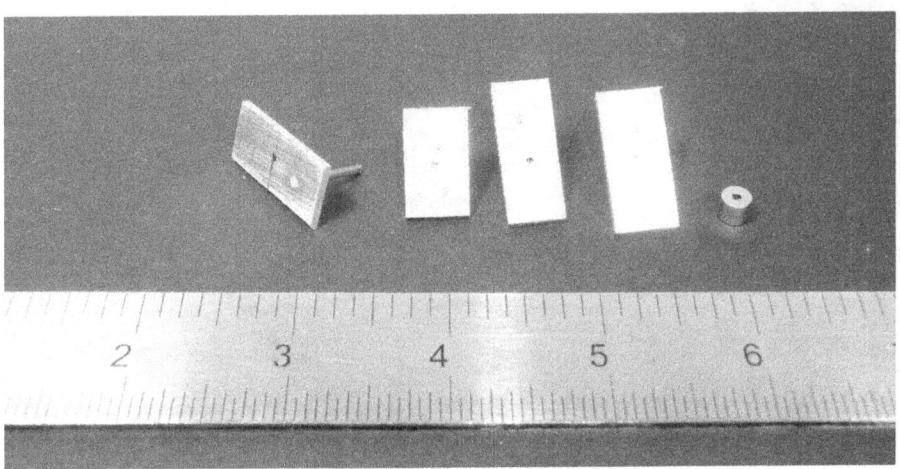

The crankshaft plate and the shaft collar are ready to be joined together with epoxy.

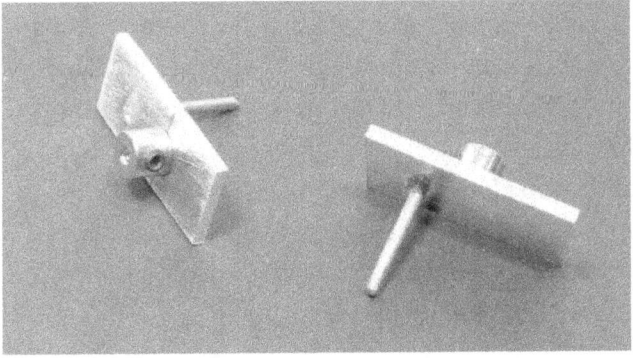

The crankshaft pin, plate, and shaft collar have been assembled.

Attach the Pedestal to the Pressure Chamber Top

The pedestal needs to be mounted so that it is aligned well with the displacer pushrod and the drive cylinder. Place a piece of straight music wire through the axle bushings to visually confirm the alignment. This will make is easier to visually check the alignment and confirm the previous marks. Glue the pedestal in place with epoxy once the correct position is confirmed.

The pedestal is attached to the pressure chamber top. There is enough room for the flywheel between the pedestal and the displacer pushrod, and the drive crankshaft pin extends over the middle of the drive cylinder.

Test the Travel Distance of the Displacer

Install the axle, flywheel, and drive crankshaft on the pedestal. Position the flywheel so that the displacer crank pin is at the bottom position of its rotation. Use a marker or a piece of tape to make a reference mark on the displacer pushrod at the point where it comes in contact with the crank pin. Now rotate the flywheel until the pin is at the top of its rotation and raise the displacer pushrod until the mark is once again even with the pin. There should be about 1/16" (1.59 mm) free space between the displacer panel and the top of the pressure chamber when the displacer crank pin is at the top of its rotation. If it appears that the displacer will be able to move up and down without impacting the top or bottom of the pressure chamber, proceed to the next step. If the displacer crank is moving the displacer too far and it is impacting the engine, correct the problem by relocating the crank pin closer to the center of the flywheel.

Trim the Displacer Pushrod

Allow the displacer panel to rest on the bottom of the pressure chamber. Measure up from the top of the pressure chamber 1 1/2" (38.1 mm) and trim the displacer pushrod at this point.

Create the Displacer Connecting Rod and Teflon Bushing

The displacer connecting rod completes the connection between the top of the displacer pushrod and the displacer crank pin on the flywheel. A flexible connection is made to the displacer pushrod with two thin pieces of duct tape. The top end of the connecting rod is a piece of Teflon tube that will slip over the displacer crank pin on the flywheel. The displacer connecting rod is made from 0.015" music wire, which is the same size as the displacer pushrod.

Cut a piece of AWG-14 heavy wall extruded Teflon tube to a length of 7/16" (11.11 mm). Use the same cutting method described earlier so that the tubing does not become deformed at the cut.

Cut a piece of 0.015" music wire to a length of at least 3 1/2" (88.9 mm) to construct the connecting rod. The exact length is not critical because it will be trimmed to fit. It may be easier to work with a longer piece and trim it to length after the bends are completed.

Place the short piece of Teflon tubing on a piece of axle stock (0.062" music wire) to help keep it straight while wrapping the connecting rod wire around the tubing. Make at least 1 1/2 wraps around the tubing with the connecting rod wire. Adjust the connecting rod wire so that it is at a 90° angle to the Teflon tube. The connecting rod wire and the Teflon tube should form the shape of a "T." Make the wraps tight enough so that they grip the Teflon tube and it does not fall out. Trim any excess wire when finished.

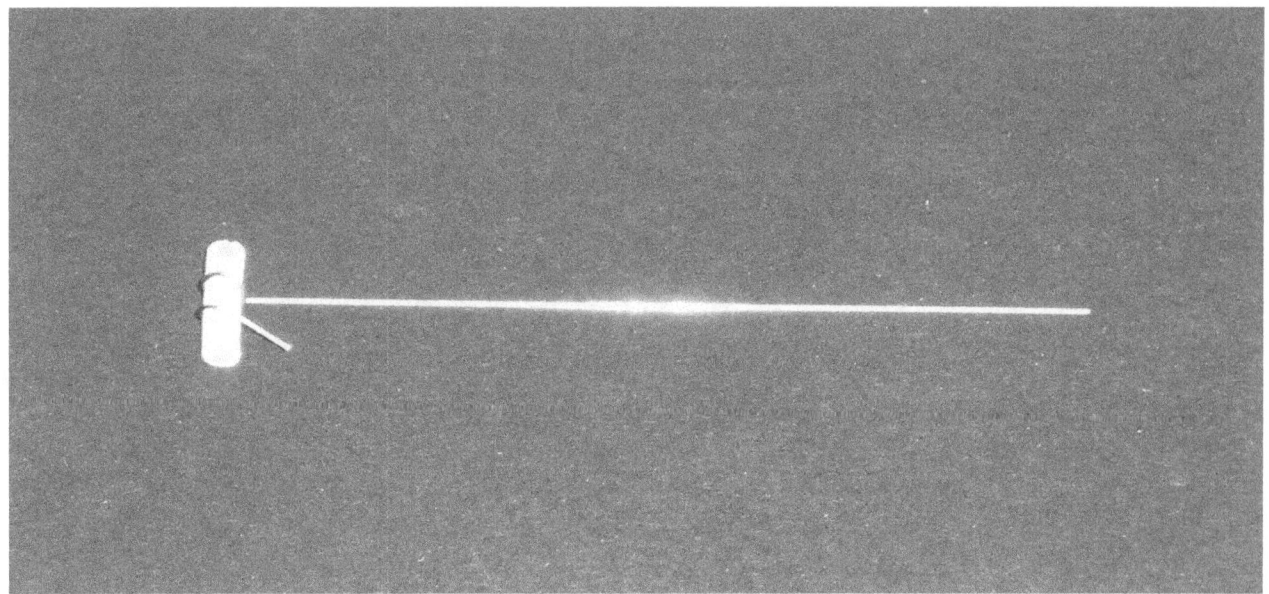

The Displacer Connecting Rod is fashioned from a length of music wire that is wrapped around a short piece of Teflon tube.

Attach the Displacer Connecting Rod

The connecting rod can now be installed between the flywheel and the displacer pushrod. Place the Teflon tube over the drive crankshaft pin on the flywheel. With the displacer panel resting on the bottom of the pressure chamber and the flywheel pin in its lowest position, trim the length of the connecting rod so that there is a gap between the end of the displacer pushrod and the end of the connecting rod. The gap should be between 1/32" (0.79 mm) and 1/16" (1.59 mm).

Cut two small pieces of duct tape to 1/8" (3.18 mm) x 1/2" (12.7 mm). Lift the displacer so that the end of the displacer pushrod is near the end of the connecting rod and fasten the two pieces together with the small pieces of duct tape. Position the duct tape so that the joint bends correctly in order to accommodate the motion of the flywheel.

Rotate the flywheel and observe the motion of the displacer panel. It should travel up and down inside the pressure chamber without touching the top or the bottom of the pressure chamber. Make any adjustments necessary so that the displacer does not come into contact with the top or bottom of the pressure chamber.

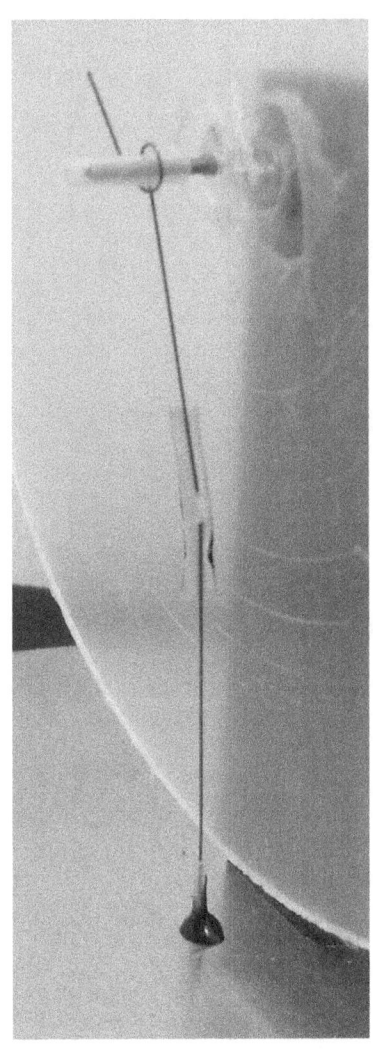

In this picture, the displacer connecting rod has been trimmed to leave a small gap between the ends of the two rods. One piece of duct tape has been applied to the joint. Note the direction of the bend at the joint and how the alignment of the tape allows it to act as a hinge.

Create the Drive Pushrod and Teflon Bushing

The drive pushrod is made just like the displacer connecting rod was made, except there is no duct tape joint in the middle of the shaft. There is a Teflon tube at the top end of the pushrod. The Teflon tube rides on the crankshaft pin. The bottom end is bent into a loop that is folded over so that it mounts flat against the drive diaphragm.

Cut a piece of AWG-14 Teflon tube to a length of 7/16" (11.11 mm). Insert a piece of 0.062" music wire inside to help hold it while wrapping a piece of 0.015" music wire around it. Make at least 1 1/2 turns around the tubing, as before. Adjust the tubing so that it is held snugly at a 90° angle to the pushrod.

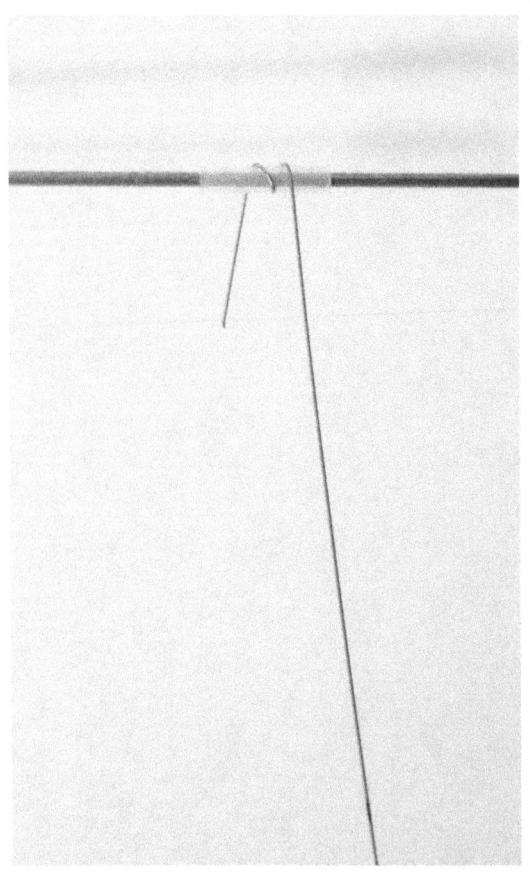

In this picture, the Teflon tube has been placed on a piece of axle wire and the pushrod wire is wrapped around the tubing to attach it to the pushrod.

Measure the distance between the center of the axle and the top of the drive cylinder. This will be the finished length of the pushrod.

Make a 90⁰ bend in the pushrod using the measurement just obtained. The distance from the axle to the top of the drive cylinder should be the same as the distance from the Teflon tube to the 90⁰ bend.

Use needle nose pliers to make a loop at the bottom of the pushrod. The loop should be about 7/16" (11.11 mm) in diameter. The loop is made in such a way so that if the connecting rod was placed on a flat surface it could stand upright with the loop flat against the table top.

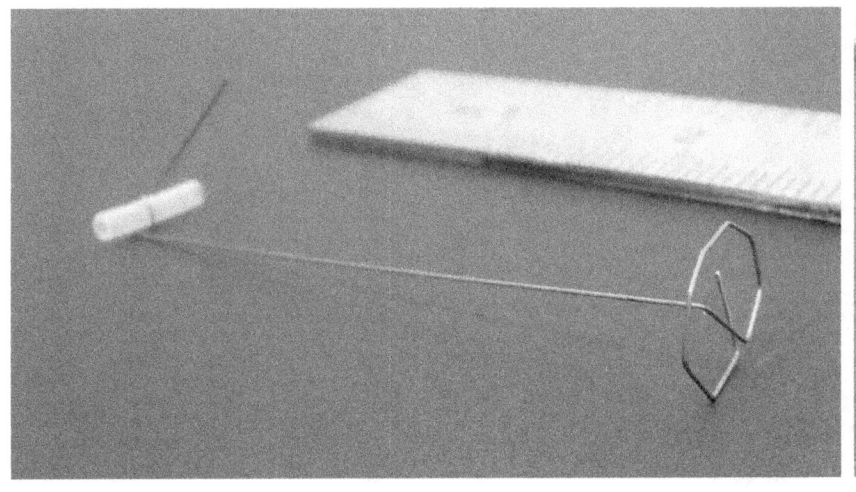 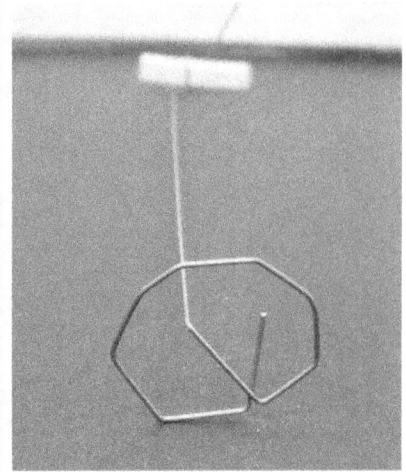

These pictures show the loop formed at the 90⁰ bend in the pushrod.

Verify the fit of the pushrod. Place it on the drive crankshaft pin and rotate the flywheel. The loop on the bottom should rise above the top of the drive cylinder at its highest point, and it should drop down into the drive cylinder at its lowest point. It should be close to even with the top of the drive cylinder when at its midpoint between high and low.

If getting the correct length is a challenge, consider making the shaft slightly longer than required, and then place a Z shaped bend near the middle of the pushrod. The Z bend can be manipulated to fine tune the length of the pushrod after it is installed.

Create the Drive Diaphragm from a Latex Glove
The drive diaphragm is made from the fingertip of a latex glove. Wash and dry a latex glove so that all the powder is removed. Cut the fingers from the glove. Stretch one of the glove fingers over the top of the drive cylinder. Pull the latex down the outside of the drive cylinder until there is only a small amount of slack near the center of the drive diaphragm.

The drive diaphragm should stay in place without any help. If it appears to be slipping, secure it by placing a rubber band around the outside of the drive cylinder. A rubber band can be made from another glove finger if necessary.

Attach the Drive Pushrod to the Drive Diaphragm
The loop at the end of the drive pushrod should rest flat against the drive diaphragm and there should be no sharp wires threatening to puncture the diaphragm. Hold the loop against the center of the diaphragm and attach it using Superglue®.

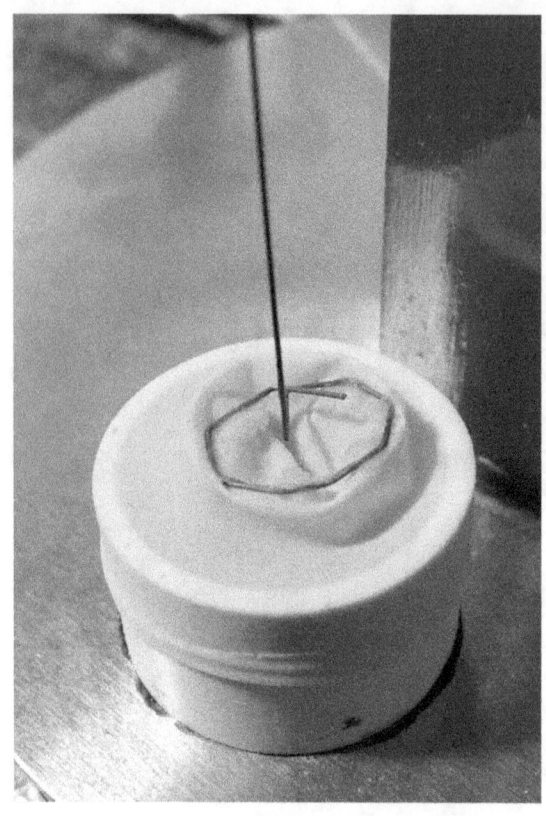

This picture shows the drive pushrod attachment to the drive diaphragm. The wire pushrod is attached to the latex diaphragm with Superglue®. The diaphragm is made from the finger tip that has been cut from a latex glove.

Set the Crankshaft Timing Angle

Adjust the flywheel and the drive crankshaft so that there is a 90° offset between the displacer crank pin and the drive crank pin. The direction of the offset will determine the direction the motor rotates when running. The drive mechanism will follow 90° behind the motion of the displacer mechanism. That means when the displacer is all the way up at the top of the rotation, the drive mechanism will be halfway up. When the displacer is all the way down, the drive mechanism will be halfway down. Use the set screws on the shaft collars to hold the flywheel and the drive crankshaft in place.

Adjust the Drive Diaphragm Tension

The drive diaphragm should be adjusted so that there is just enough slack in the latex to allow the engine to rotate without stretching the material. If the material is too tight the engine will not run well because extra energy will be required to stretch the diaphragm. If the diaphragm is too loose the engine will not run well because the loose diaphragm will inflate and deflate without causing the crankshaft to move. Adjusting the tension of the drive diaphragm is one of the adjustments that can be made to fine tune the performance of the motor.

Check all the Connections

The engine should now be fully assembled and ready for its first run. Check all the connections by rotating the flywheel slowly and observing all the moving parts. Nothing should be falling apart when the flywheel is rotated. If the connecting rod or the pushrod becomes disconnected during this test, make adjustments so that they do not fall off.

Observe the motion of the displacer panel inside the pressure chamber and ensure that it does not impact the top or the bottom of the pressure chamber. It should move without any obstruction.

Run the Engine!

This Stirling Engine should run well over hot water. Fill a coffee cup or similar container with near-boiling water. Place the pressure chamber on top the cup of hot water. Allow it to warm up for 10 to 20 seconds. Turn the flywheel to start the engine.

The motor will continue to run as long as there is a temperature differential of 20° F (11° C) (or more) between the top and bottom surfaces of the pressure chamber. It may be possible to fine-tune the engine to operate on an even lower temperature differential with a little care and patience.

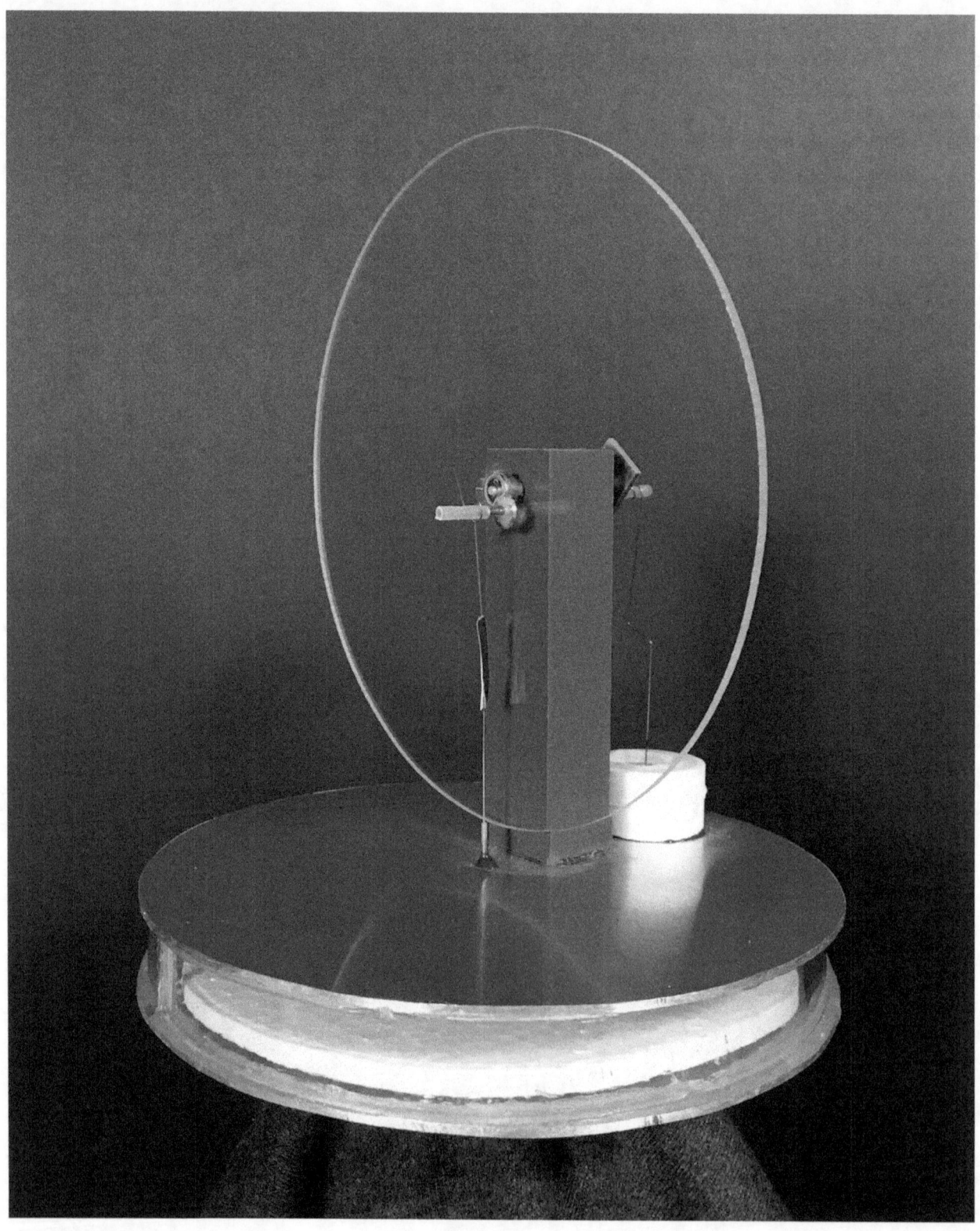

The finished 6" (15.2 cm) round engine.

Trouble Shooting Tips

If the engine is not running well, it may be because of a problem in one of these four areas:

Temperature Differential: As mentioned previously, the engine should run with a temperature differential of 20° F (11° C). Setting the engine on a cup of near-boiling water in a 70° F (21° C) room should provide a temperature differential of about 100° F (56° C). If the engine is not running under these conditions, there are one or more other problems that will need to be fixed, such as a small pressure leak or friction.

It may be possible to overcome a small pressure leak or friction by increasing the temperature differential. This will increase the power output of the engine and possibly overcome a small amount of friction or a small pressure leak.

Increase the temperature differential by adding ice to the top of the motor while the bottom is being warmed by the heat source. Do not attempt to add more heat, as this can damage the engine. The Styrofoam displacer material may melt if the heat source is too hot.

Pressure Leaks: It does not take much of a leak to prevent the engine from running well. There are a couple of ways to test for a pressure leak. The first method is to observe the behavior of the diaphragm when the engine is at running temperature.

Disconnect the drive diaphragm pushrod from the crank pin. Place the engine on a cup of hot water and wait a few moments for the bottom side to heat up. Now, rotate the flywheel so that the displacer rises and falls inside the pressure chamber and observe the motion of the drive diaphragm. The diaphragm should move up and down in response to the heating and cooling of the air inside the pressure chamber. If this motion is not present, or if it is very limited, there may be a pressure leak. It may also be possible that the tension of the diaphragm is too tight or too loose.

The other method for leak testing also involves removing the diaphragm pushrod from the crank pin. Once it is disconnected, pull upward on the pushrod for 5 to 10 seconds to inflate the diaphragm. Release the pushrod and observe the diaphragm. If it immediately deflates and returns to a low or neutral position, there may be a pressure leak.

Pressure leaks can happen at a number of places:

- Holes in the drive diaphragm
- Leaking around the edge of the drive diaphragm
- Leaking glue joints in the pressure chamber
- Leaking through excessive clearance around the displacer pushrod.

It may take a bit of detective work to find the leak. It may be possible to patch a small leak with a small drop of glue or silicone sealant, or it may be necessary to replace the defective part.

Friction: Small amounts of friction can have a huge impact on an LTD Stirling engine's ability to run. Friction occurs at every point where two moving parts touch, and at every point where a moving part contacts the atmosphere. In the micro-horsepower world of LTD Stirling engines, a tiny bit of friction can stop the engine from performing.

Check the rotation of the axle by removing the connections to the displacer and the drive diaphragm and spinning the flywheel. The flywheel should coast to a stop after about 30 seconds after receiving a good spin by hand. If the flywheel does not spin freely there is a problem with friction somewhere in the axle assembly. Locate the cause of the friction and repair the problem.

The flywheel rotation should be smooth and silent during the spin test. Vibration and noise are both indications of friction.

Crankshaft Timing: There must be a 90^0 phase difference between the two crank pins. This means that when one pin is in the 12:00 o'clock position, the other one is at either 9:00 o'clock or at 3:00 o'clock. The engine will run with a phase difference in either direction. The only difference will be the direction the engine rotates while running.

When the bottom of the engine is heated, the engine will run with the motion of the displacer moving ahead of the drive diaphragm. This means that when the displacer is at the top of the pressure chamber, the drive diaphragm is halfway up and moving in an upwards direction. As the flywheel rotates and the displacer comes to the lowest point in its travel, the drive diaphragm is halfway down and moving downward. It is important to know which way the engine will run so that the initial push to get it started is in the same direction.

Maintenance and Operating Tips

Do not lubricate the engine. Adding oil to the Teflon tube bearings or gland will increase the friction in your motor. If you sense that there is friction at one of these points, it is probably caused by poor fitting or poorly aligned parts.

Be prepared to replace the latex diaphragm. The latex lasts about a year under light use. The material will eventually weaken from exposure to the heat and the elements and will need to be replaced. It only takes a few minutes to replace a torn diaphragm.

Avoid excessive heat. Styrofoam, acrylic, and some adhesives do not tolerate exposure to high temperature well.

Try adding helium for improved performance. Filling the pressure chamber with helium can boost the engine performance by as much as 15%. These engines are not sealed, so helium will not last long. Vent tubes can be added to the pressure chamber to facilitate adding helium. Helium filling is shown in detail in the previous Jim Larsen book, "Three LTD Engines You Can Build Without a Machine Shop."

The engines pictured in this collection have successfully run from the heat of a warm hand in a cool environment. Running a handmade Stirling engine in this manner will require great patience and attention to detail. Since all parts are hand made by the builder, every engine will be slightly different, and results will vary with each one. The key is to watch for friction, pressure leaks, balance issues, and temperature differential.

Alternative Construction Options

One of the joys of engineering is adapting a design to make it work. That can also be true with these Stirling engines. Every effort was made to use products that are widely available in the US and elsewhere. That doesn't mean that the parts and materials will be readily available everywhere. Here are a few alternatives that can be explored as options for these engines.

Alternative Options for Displacers

Closed Cell Foam Insulation Board
Closed cell foam insulation board can be used as an alternative to the white Styrofoam displacer. Closed cell foam insulation board is a construction supply used to insulate foundations during home construction in the US. This material is heavier and stiffer than Styrofoam, and tolerates heat a little better. It can still be cut with a hot wire foam cutter, or with a conventional saw blade.

Balsa Wood
Balsa wood is light in weight and tolerates heat very well. It can be found in stores that sell model airplane supplies. Light weight balsa wood can be purchased in sheet form and it can be cut to shape with a sharp utility knife.

Paper
This technique will require some patience. Build a hollow paper structure that is the same dimensions as a foam displacer. Use glue that will tolerate heat well. The hollow form will be very light weight.

Cardboard
The displacer can be cut from a flat piece of cardboard that has been cut from a cardboard box. The holes in the end of the corrugated sheet do not need to be blocked or filled. This material is likely to be heavier than Styrofoam, but will be easier to locate and easier to work with.

Foam Core Poster Board
Foam core poster board is available in stores that sell art supplies, or in shops that make and sell picture frames. It is available in a variety of thicknesses. It is a sheet of Styrofoam with paper laminated to both surfaces. It weighs more than plain Styrofoam, but it does not require a hot wire foam cutter to create the sheets used to make the displacer.

Alternative Options for Drive Cylinders
The drive cylinders on these engines are made from a short piece of PVC plastic pipe that is a standard building material in the US, but is not as easy to find in other parts of the world. There are many kinds of materials that can be used as alternatives. The dimensions do not have to be an exact match, but any alternative will still have to fit in the space available on the top plate of the engine. Here is a quick brainstorm of some alternative materials that can be used for the drive cylinder:

- 35 mm Film Can
- Pill bottle
- Plastic Tube
- Small Pipe (metal or plastic)

Alternative Options for Drive Diaphragms

Helium Quality Balloon
Helium quality party balloons make excellent drive diaphragms.

Nitrile Rubber Glove
Nitrile rubber gloves are becoming more common as an alternative to latex gloves as a result of latex allergies. Nitrile is not as flexible as latex, but with care and patience it can be made to work.

Airpot Actuator
Airpot Actuators are complete assemblies with a glass cylinder and a graphite piston. These can be adapted to become an almost friction-free drive mechanism for an LTD Stirling engine. They will probably have to be ordered online, and are likely to double the cost of any project in this book.

Alternative Options for Pressure Chamber Sidewalls
Clear acrylic is the first choice for sidewall material because it allows us to see the inner workings of the engine, but transparency is not a requirement. These engines will work with many types of material as a sidewall. The properties required are that it be a decent thermal insulator, and that it is not so porous as to cause a pressure leak. Here are some alternatives for sidewall material:

- Wood
- 3D printed parts
- 4" PVC pipe
- Stiff cardboard

Alternative Options for Flywheels
- Music or Data CD
- Cardboard

Build a Hot Wire Foam Cutter

A hot wire foam cutter is a very effective method to cut a piece of Styrofoam into the exact dimensions needed for a displacer. Foam cutters can be purchased online and at some hobby and craft stores, or one can be made with a salvaged power supply, some Nichrome wire, and a simple wood or plastic frame.

Nichrome Wire
According to Wikipedia, "Nichrome is a non-magnetic alloy of nickel, chromium, and often iron, usually used as a resistance wire." "Resistance wire" is wire that becomes hot when electrical current is passed through it. The wire offers some resistance. The resistance produces heat and also prevents the power supply from overloading due to a short circuit.

Nichrome wire is available on Amazon.com for a few dollars. At the time of this writing they carried a product called, "Woodland Science Foam Cutter Replacement Wire." A package containing 4 feet (1.2 meters) of wire is priced under $3.00 (US). Searching the Internet will provide a long list of retailers who sell Nichrome wire.

Crimp-on ring terminals are attached to each end of the Nichrome wire. These are attached to a nut and bolt arrangement on each side of the framework. The crimp-on ring terminals are not necessary, but they do simplify the assembly of the cutting wire to the frame, and they extend the life of the wire. The alternative method is to wrap the wire around the small bolt and hold it in place with a nut.

Power Supply

An ideal power supply is one with a variable voltage output of up to 16 volts with a relatively low amperage rating. Variable power output is an important feature. This allows for the adjustment of the power output of the power supply to match the load of the resistance wire and control the temperature. If the power is too high the wire will overheat and break easily. If the power is too low, the wire will not be hot enough to cut the foam. Different types of foam will have different melting points and the temperature of the cutter may need to be raised or lowered depending upon the material chosen. The home builder has several possible solutions for a variable output power supply:

1. **Model Train Transformer** – Model train transformers control the speed of the toy train by changing the power output of the transformer. Voltage can be controlled in a range of 0 to 16 volts. Amperage varies depending upon the size of the train set being powered.
2. **Automotive Battery Charger** – This can be used on a low setting (i.e., trickle charging) with the addition of a household dimmer switch or similar rheostat for adjusting the power output of the battery charger.
3. **AC Adapters** – Adapters from small electronic devices will sometimes work as a power supply for a hot wire foam cutter. This is not the best choice because the power output cannot be controlled and some small units do not have enough power to heat the wire. But with a little luck and the right Nichrome wire it will work. It may be necessary to try several different small power supplies before one is found that is a good match for the power requirements of the selected wire.

The power supply must be capable of withstanding the large load caused by the Nichrome wire. The power supply and the connecting wires must remain cool during operation in order to be safe. If overheating is detected, stop immediately and disconnect the power supply from the AC power.

A piece of two conductor electrical cord (such as lamp cord) is used to connect the power supply to the cutting wire. This wire needs to be heavier than the cutting wire to prevent overheating. An old power cord from a 110 volt electrical appliance will work well.

Framework

A framework must be built to hold the cutting wire taught, and it should be made of a non-conductive material so that it does not create an electrical short circuit. The framework described here is made of wood.

Nichrome wire expands when it heats. There is enough expansion to cause the wire to go slack when it is heated up enough to cut foam. When the wire becomes slack it can become difficult to make a smooth cut in the foam. The design illustrated here uses an elastic shock cord to keep the wire under tension. This mechanism keeps the wire under tension at all times and makes it possible to achieve strait smooth cuts.

The frame needs to be wide enough to accommodate material up to 6" (15 cm) wide. The frame shown here has 7" (17.8 cm) clearance between the arms.

The frame is made from dry wood. Wood that is green or wet will conduct electricity and cause problems.

Controlling Depth of Cut

The depth of cut is controlled by setting small spacer blocks on the work surface and letting the points of the wood frame rest on the spacers. The foam is then held against the work surface slid through the frame. The hot wire will slice through the foam. Very little pressure is required. Creating multiple sets of spacers of various thicknesses will allow for adjustments to the thickness of the foam slab. The thickness adjustment can be fine-tuned by adding a few playing cards under each spacer.

The variable voltage power supply should be adjusted to the lowest setting that still provides enough heat to melt the foam. Overheating the wire is not necessary. Start with a low setting and practice on a piece of foam. Increase the power if necessary.

James Senft recommends cutting the displacer out of the core of the foam. He states in his book, "An Introduction to Low Temperature Differential Stirling Engines" (James R. Senft, Moriya Press, 1996) that if one side of the foam is left as the "factory finish" and the other side is cut, the foam can warp and cause friction inside the pressure chamber. This warping is prevented if both surfaces (top and bottom) of the displacer are cut with the hot wire foam cutter.

Foam for these projects can be found from several sources. It may be salvaged from packing material, or purchased in large slabs from the insulation department at a building supply store.

Some types of foam will release harmful gasses when they are heated. Always do foam cutting in a well-ventilated area, especially when unsure about the chemical makeup of the foam.

The hot wire foam cutter gets hot enough to burn skin, and can cause a fire hazard if not used properly. Always use it with extreme caution and unplug the power supply from the wall between uses.

Use Extreme Caution

This cutting mechanism creates heat by making a short circuit across an electrical power source. There is a risk for fire, burns, electrical shock, or electrocution. You, as the builder, are responsible for your own safety. Take appropriate safety measures to prevent injury and to prevent fire. Discontinue use at the first sign of trouble. The author makes no promises that this is foolproof or safe. Always work in a well ventilated area that is free of combustible materials. Never leave power supplies plugged in or unattended. Never work with equipment that is malfunctioning. Always work in a manner to keep yourself and those around you safe.

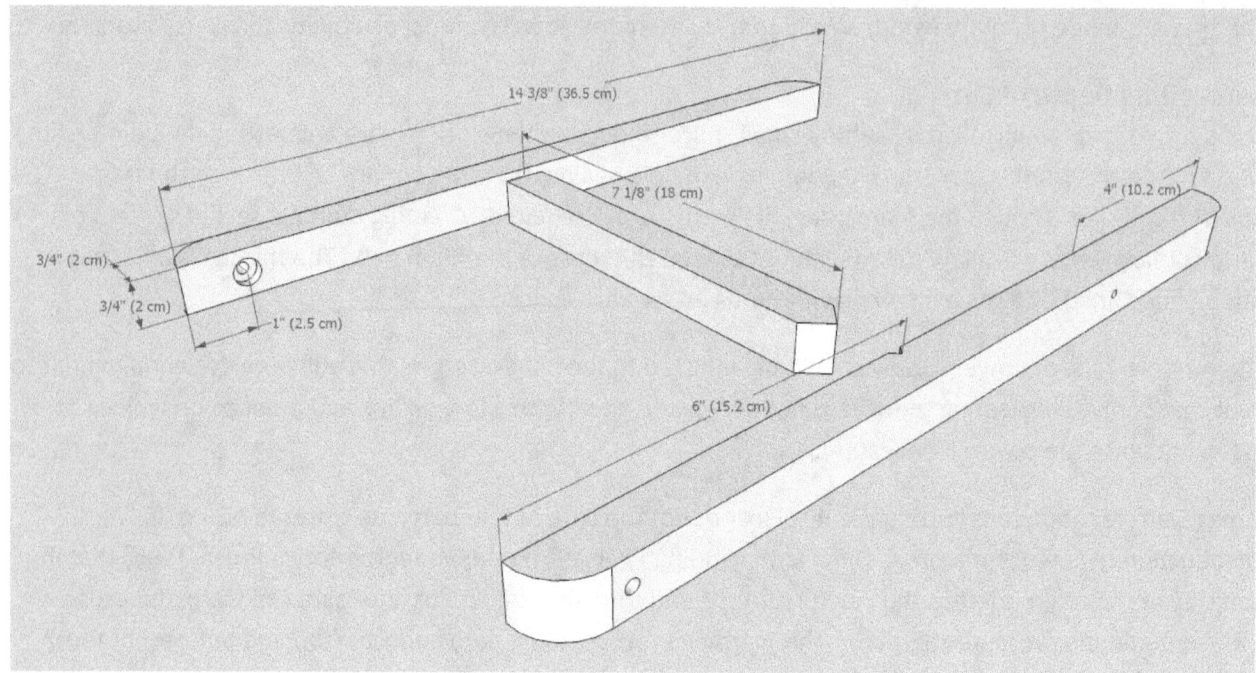

This image shows the dimensions of the framework for a hot wire foam cutter.

Construction Steps

The wood frame consists of a left arm, a right arm, and a middle arm. All parts are made of wood.

The left and right arms (shown above) are 14 3/8" (36.5 cm) long and are cut from 3/4" (2 cm) square stock.

The middle arm is 7 1/8" long and is also cut from 3/4" (2 cm) square stock.

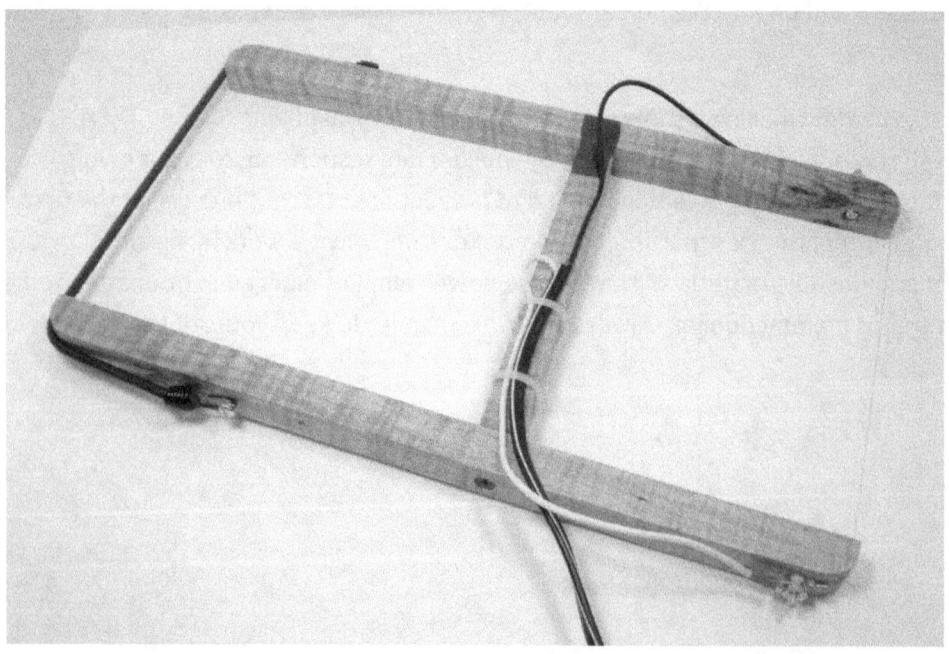

This is the assembled hot wire foam cutter. The Nichrome wire is on the right side. The bungee cord on the left holds tension on the frame to keep the Nichrome wire tight.

1. Cut each piece to length.
2. Drill and countersink a hole 1" (2.5 cm) from the bottom end of the left and right arm. This hole will hold a small bolt and nut for mounting the Nichrome wire. The countersink is on the inside of the frame, and prevents the head of the mounting bolt from obstructing the cutting area. Choose a drill bit that offers a good fit for the bolt used to attach the wires.
3. Cut a point on one end of the middle arm. This point will ride in the notch to be cut in the right arm.
4. Cut a small notch (1/8" (3.2 mm) deep) in the right arm, at a point 6" (15.2 cm) from the bottom end of the arm.
5. Round the outside corners on the top and bottom of both arms.
6. Cut a small notch in the bottom of each arm to hold the Nichrome wire.
7. Attach the middle arm to the left arm at a point 6" (15.2 cm) up from bottom of the arm. The pointed end of the middle arm must be aligned with the notch in the right arm. Attach the middle arm to the left arm with wood glue. This joint may be reinforced with a wood screw if necessary.
8. Place the pointed end of the middle arm into the notch of the right arm. Heavy tape can be added to the joint to help keep it aligned.
9. Insert small bolts into the holes at the bottom of each arm.
10. Cut a piece of Nichrome wire to about 12" (30.5 cm) in length. Attach a crimp-on ring terminal to each end of the wire. For added strength pass the wire up through the crimp of the ring terminal, loop it around the terminal, and then pass it back down through the crimp. Twist the wire a few times before crimping. The wire needs to be long enough to allow 7" (17.8 cm) between the ends of the wood frame when the wires are attached to the bolts. Attach the wire to each bolt with a small nut.
11. Attach a small bungee cord to the top of the wood frame to create tension on the wire and hold the frame together.
12. Attach wire leads to each bolt and fasten with a second small nut. Tie or tape the lead wire to the frame to prevent stress on the attachment point.
13. Create some wooden spacers to control the depth of cut. Cut the spacers to the same thickness that the foam is to be cut. To make a foam slab that is 1/4" thick, cut spacers that are 1/4" thick.

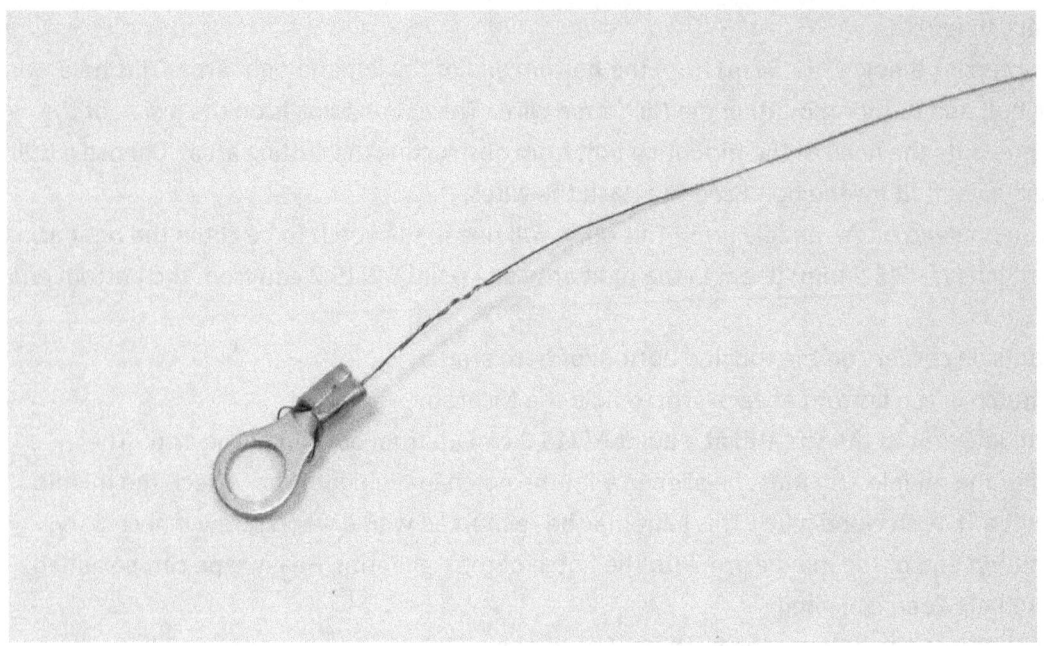

Fasten an electrical connector to each end of the Nichrome wire.

This flexible joint is necessary because the Nichrome wire grows in length when it is heated. The frame compensates for this and keeps the wire tight.

This image shows a close up of the Nichrome wire attached to one conductor.

If the cutting wire is being held tight by the framework, and if the lead wires are attached to each end of the cutting wire, the next step is to attach the power supply and begin a test cut.

WARNING: The cutting wire gets hot and can cause injury. Use with extreme caution at all times. Monitor the power supply and connecting wires constantly and disconnect electrical power at the first sign of overheating.

Use a low power setting for the first test cuts. Adjust the power setting until the wire is hot enough to cut through the foam.   Overheating the cutting wire will shorten the life of the wire and cause it to break prematurely.

Place the spacers on the work surface and set the ends of the hot wire foam cutter on the spacers. Turn on the power. Slowly slide a block of foam through the hot wire to create a straight cut. Repeat the process with a second cut. Thin slabs of foam are less likely to warp if they have a hot wire cut on both sides.

The finished slab of foam will be the same thickness as the spacers. The spacers hold the hot wire at a measured distance above the work surface.

The small block of wood controls depth of cut. Slide a block of foam through the cutter at this point to create a slab of foam. The foam will be the same thickness as the shim.

Making the Best Bearings: Friction Research

When working in the micro-horsepower world of small Stirling engines it is important to do everything possible to reduce friction. And when one is trying to keep the project as economical as possible, it is also important to find affordable solutions.

This research began when I decided to validate my assumption that a plastic material known as UHMW polyethylene really was a good choice for making small bearings for Stirling engines. Everything I was reading about UHMW polyethylene supported the idea that it had a very low coefficient of friction and excellent wear characteristics that would make it almost as good as ball bearings. Part of the description found at Wikipedia (at the time of this writing) reads:

> "Ultra-high-molecular-weight polyethylene (UHMWPE, UHMW) is a subset of the thermoplastic polyethylene. Also known as high-modulus polyethylene, (HMPE), or high-performance polyethylene (HPPE), it has extremely long chains, with a molecular weight usually between 2 and 6 million. The longer chain serves to transfer load more effectively to the polymer backbone by strengthening intermolecular interactions. This results in a very tough material, with the highest impact strength of any thermoplastic presently made.

> "UHMWPE is odorless, tasteless, and nontoxic. It is highly resistant to corrosive chemicals except oxidizing acids; has extremely low moisture absorption and a very low coefficient of friction; is self-lubricating; and is highly resistant to abrasion, in some forms being 15 times more resistant to abrasion than carbon steel. Its coefficient of friction is significantly lower than that of nylon and acetal, and is comparable to that of polytetrafluoroethylene (PTFE, Teflon), but UHMWPE has better abrasion resistance than PTFE."

I had been using UHMW in several projects prior to this research because if the impressive characteristics of a low coefficient of friction, high resistance to abrasion, and self-lubrication. The friction testing done for this book has changed my mind. UMHW did not finish first in the testing. There are some additional characteristics of this material that make it hard to work with, and these contributed to the less-than-perfect scoring in the testing.

It is difficult to drill smooth, accurately sized holes in UHMW. As a combination of the elasticity of the material and its resistance to abrasion, the material compresses around the drill bit, causing the drill to act similar to a nail in wood. After the drill bit is removed, the compressed material around the hole slowly returns to its original shape, resulting in a hole that is smaller than the drill bit used to make it. The hole will continue to shrink for the next several hours. This phenomenon caused the testing to show shorter and shorter run times for the UHMW if the holes were not re-drilled before the subsequent tests.

The surface of UHMW is very rough after being cut with a drill or a saw. Since the material is highly abrasion-resistant, it is difficult to smooth these rough surfaces, and rough surfaces do not make good bearings. It may be possible to use heat to either drill the holes or polish the interior surface of the holes. That solution was not attempted or tested here. I did use polishing compound on the inside of the holes and found that it improved the run times in testing, but after 15 minutes of polishing the material was still significantly slower than the first place winner.

And the Winner Is...

Of all the materials tested, the best performing material in the flywheel glide testing was polytetrafluoroethylene (PTFE). Some will know it better by its brand name, Teflon. PTFE has one of the lowest coefficients of friction of any solid. It is manufactured in many forms, including accurately sized extruded tubing that makes it very easy to use as bushing material for making small bearings.

One Other Surprising Result

I competed in the Soap Box Derby as a young teenager. On one race day, the parent of one of the competitors brought a big box of little oil cans and made them available to any of the young men who wanted to lubricate the ball bearings in their wheels. Most of the excited young drivers scrambled to grab an oil can and lubricate their bearings. Everyone wanted any advantage they could get to help them speed down the track. This was a sly trick, as the addition of the oil actually makes the ball bearings harder to turn. This crafty father didn't let his own son lubricate the bearings on his car. His car came in first place.

So many people assume that lubricating bearings is a good thing that no one was suspicious of this Good Samaritan's gifts to the competitors of the race. A good friend warned me not to put the oil on my bearings, but it was already too late. I can't blame the oil in the bearings, but I can say that I did not win that day. And my good friend who did not oil his bearings – he won second place!

The flywheel glide testing that was performed for the research in this book confirmed what that Soap Box Derby parent knew to be true. Lubricating the bearing surfaces with light 3-in-1 oil or with silicone lubricant resulted in decreased glide times on every surface tested. All the bearing materials provided better performance when they were not lubricated.

Testing Apparatus

A flywheel was created using a CD disk with a piece of 0.0625" (1.5875 mm) music wire as the axle. A small wooden spool was attached to the axle near the flywheel. A small weight was attached to a length of string. The string was wound around the spool on the flywheel to a predetermined point and the weight was then dropped. Using the weight and string to start the flywheel regulated the tests so that the flywheel would be spun with the same amount of force in every trial. The spinning flywheel was timed from the start of motion until it stopped moving. Ten trials were performed in each test cycle and the glide times of those ten trials were averaged.

Three test beds were crafted with a wood base and a variety of bearing materials. Holes were drilled in the bearing material to match the size of the axle. Holes were drilled with a 1/16" (1.59 mm) drill bit, and with a 5/64" (1.98 mm) drill bit. Softer materials (like wood) did not allow the axle to spin freely with the smaller hole. Because of this, the larger drill bit was used when the smaller hole proved ineffective.

Materials were chosen because of their availability for testing and the belief that they would also be readily available to most of the people building one of the Stirling engines in this book.

Test bed #1 was used to test the following materials:

1. Steel can (a metal can from the kitchen that originally held canned chicken)
2. UHMW Plastic, thickness: 3/8" (9.5 mm)
3. Hardwood (Mahogany), thickness: 3/4" (19 mm)
4. Soft wood (Pine), thickness: 3/4" (19 mm)
5. Schedule 40 white plastic PVC pipe, thickness: 1/8" (3.18 mm)
6. Clear Acrylic Sheet, thickness 1/4" (6.35 mm)
7. Aluminum, thickness: 0.08" (2 mm)

The test results from this first collection of materials showed that the thickness of the material had more impact on the glide times than the material itself. While this was an important lesson, it was not the kind of results being sought after. The intent of these tests was to see which material performed best as a bearing for a Stirling engine. In order to find that out, the materials used in the testing would have to be of a similar thickness, with a similar pattern of contact against the axle of the flywheel.

Test bed #2 was made using materials of similar thicknesses. All of the bearing materials for test bed #2 were between 1/16" (1.59 mm) and 1/8" (3.18 mm) thick.

Test bed #2 was constructed with the following bearing materials:

1. Clear Acrylic sheet
2. UHMW polyethylene
3. Hardwood (maple)
4. Schedule 40 white plastic PVC pipe
5. Brass wire
6. Steel flat washer
7. Steel wire
8. Hardwood (English walnut)
9. Hardwood (Padauk)

The soft pine wood performed very poorly in the first tests and was not included in subsequent tests. A variety of hardwoods were chosen to see if there would be any significant differences in wood type.

Test bed #3 was built specifically for testing PTFE (Teflon tube) as a bearing material. The tubing was cut to a length of 0.18" (4.58 mm) and held in a piece of hardwood that was about 1/8" (3.18 mm) thick.

One round of testing was done on test bed #3 with the music wire axle wire running in the Teflon tube. After those tests were completed, the axle was covered with a heat-shrink PTFE Teflon tube and the tests were repeated with the Teflon coated axle in the Teflon bearings. A variety of configurations were tried in order to find the best combination of diameters between the heat-shrink tubing and the Teflon tube bearing material.

Testing Procedure

There were a minimum of 10 trials with each material in each state of lubrication:

1. Dry
2. Lubricated with Silicone lubricant
3. Lubricated with 3-In-1 oil

All testing of PTFE (Teflon) was done without silicone or 3-In-1 oil. No lubricants were added.

If times were trending upward for a particular material it was assumed that performance was improving as a result of run-in. Sets of 10 trials each were repeated until the times stabilized. The set of 10 trials with the best average time is the one used for comparison in the results.

The UHMW polyethylene was the only material that showed a decrease in performance over time. This was determined to be caused by the shrinking size of the hole causing the axle to bind. A drill bit was passed through the holes again to resize them for a better fit. The best times were used for comparison. (The UHMW material had to have the holes resized once about 24 hours after they were first drilled and then again about 48 hours after the initial drilling.)

After the holes in the UHMW polyethylene stabilized and stopped binding on the axle the glide times showed a steady increase in performance as trials continued. This indicates that the UHMW material gets better over time as the bearing surfaces are run-in. The UHMW bearing holes were polished with polishing compound to smooth the bearing surface and the testing was repeated.

All of the dry (non-lubricated) tests were completed before any lubricants were used. The first lubricant tested was spray silicone. All surfaces were tested with the silicone lubricant and then the axles and bearings were cleaned before being tested with 3-In1 oil. The axle and the bearing surfaces were wiped down with a paper towel and cleaned with alcohol to remove the silicone lubricant before the oil was applied.

The flywheel rotation was started by dropping a weight attached to a string wound around a bobbin on the axle, near the flywheel. The end of the string was looped over a small pin on the bobbin to help ensure that the energy applied to the flywheel was consistent during all trials.

It was determined that a sample size of 10 was sufficient. The average time remained relatively stable after 8 or 9 trials.

The materials of test bed #1 that were more than 1/8" (3.18 mm) thick were not tested after the initial round. The Aluminum and the steel can were kept in the testing because they were similar in thickness to the other materials being tested.

Recommendations

It is assumed that the best flywheel arrangement with the least amount of friction will be produced in a machine shop with accurately fitted bearings that have been designed specifically for the loads present in the engine being built. What would be a good substitute for home builders who don't have a machine shop and are building their engines by hand? The following recommendations are based on the results of these tests.

Teflon Tube (PTFE)

The Teflon tube had the best performance and was the stand out winner in the glide time testing. While most of the materials tested produced glide times between 18 seconds and 25 seconds, the Teflon tube bearings produced glide times averaging 63.5 seconds. These results were achieved with no added lubrication. Coating the steel axle with heat shrink Teflon proved less effective due to the uneven thickness of the heat shrink tubing after heat was applied.

Pros:

1. Best glide times of any of the materials tested.
2. Easy to work with.

Cons:

1. It is a special order item not usually on the shelf in local stores.
2. There is an added expense. However, a $12 (US) investment will yield enough PTFE to provide a lifetime supply of small bushings for the average Stirling engine builder.

Ultra High Molecular Weight Polyethylene (UHMW, UHMWPE)

UHMW polyethylene performed nearly as well as the Teflon tube, but required a significant amount of work in order to achieve those results. In order to make the material thin enough to be effective it had to be sliced on a band saw. The holes had to be re-drilled several times to overcome the material's tendency to "heal" the holes. And the interior of the drilled holes must be polished smooth. When UHMW is simply drilled and installed without taking these extra steps it performs no better than a bearing made of wood.

Pros:

1. Very good glide time results are possible.
2. UHMW can often be found for sale at woodworking stores, as it is commonly used to make jigs and fences for power tools.
3. Good resistance to abrasion creates long wearing bearing surfaces.

Cons:

1. UHMW comes in thick slabs that require power tools and extra effort to cut it down to the dimensions suitable for these projects.

2. The interior of the holes must be polished in order to achieve good results. Unpolished UHMW performed no better than simple wood bearings.
3. Holes drilled in UHMW tend to heal themselves for a period of time and must be re-drilled. If a hole is re-drilled it must also be re-polished.
4. There is an added expense factor. A 24" (61 cm) x 4" (10 cm) x 3/8" (9.5 mm) sheet of UHMW sells for between $25 and $30 (US). However, a single sheet that size is more than enough to build all the projects in this book, and then some.

Steel Flat Washer

The steel flat washer was the third best performing material in these tests. While it did not perform as well as either Teflon or UHMW polyethylene, it outperformed all of the other materials tested. While most common materials provided glide times of 20 to 25 seconds, the bearings made by drilling holes through a steel flat washer yielded an average glide time of 35.65 seconds. The flat steel washers used in this test are 1" (25 mm) in diameter with a 3/8" (9.5 mm) bolt hole. The holes for the axle bearing were drilled through the flat metal using a drill press and a 5/64" (1.98 mm) drill bit.

Pros:

1. Good glide time test results.
2. Flat washers are very affordable. They can be purchased at hardware stores for a few pennies each or can often be salvaged from other sources for free.
3. Easy to work with.

Cons:

1. Drilling the holes requires an electric drill (hand held or a drill press) and a method for holding the small washers securely while being drilled. This is not a big problem, as a drill is needed for other operations in this project.
2. Two other materials performed better in testing.

By the Numbers

The first chart shows the glide times in seconds for the non-lubricated testing. The bare steel axle in a Teflon (PTFE) bushing had the longest glide time. The other tests with Teflon involved Teflon coatings on the axle. The "Large Tube" and "Medium Tube" labels are references to the thickness of the heat-shrink tubing that was applied to the axle.

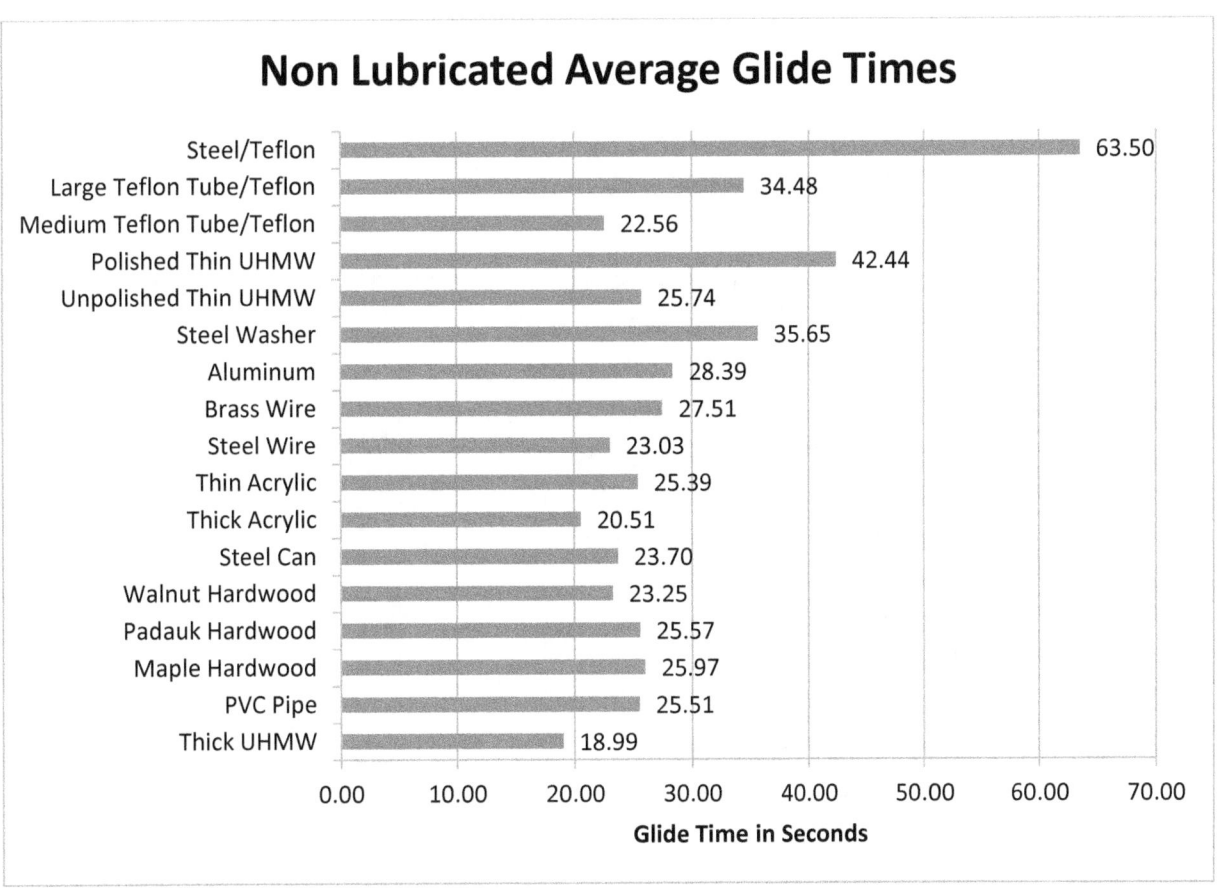

Non Lubricated Average Glide Times

Material	Glide Time in Seconds
Steel/Teflon	63.50
Large Teflon Tube/Teflon	34.48
Medium Teflon Tube/Teflon	22.56
Polished Thin UHMW	42.44
Unpolished Thin UHMW	25.74
Steel Washer	35.65
Aluminum	28.39
Brass Wire	27.51
Steel Wire	23.03
Thin Acrylic	25.39
Thick Acrylic	20.51
Steel Can	23.70
Walnut Hardwood	23.25
Padauk Hardwood	25.57
Maple Hardwood	25.97
PVC Pipe	25.51
Thick UHMW	18.99

The next chart provides a comparison between the glide times for various materials when lubricants were added. Adding lubricants decreased the glide times for every material tested.

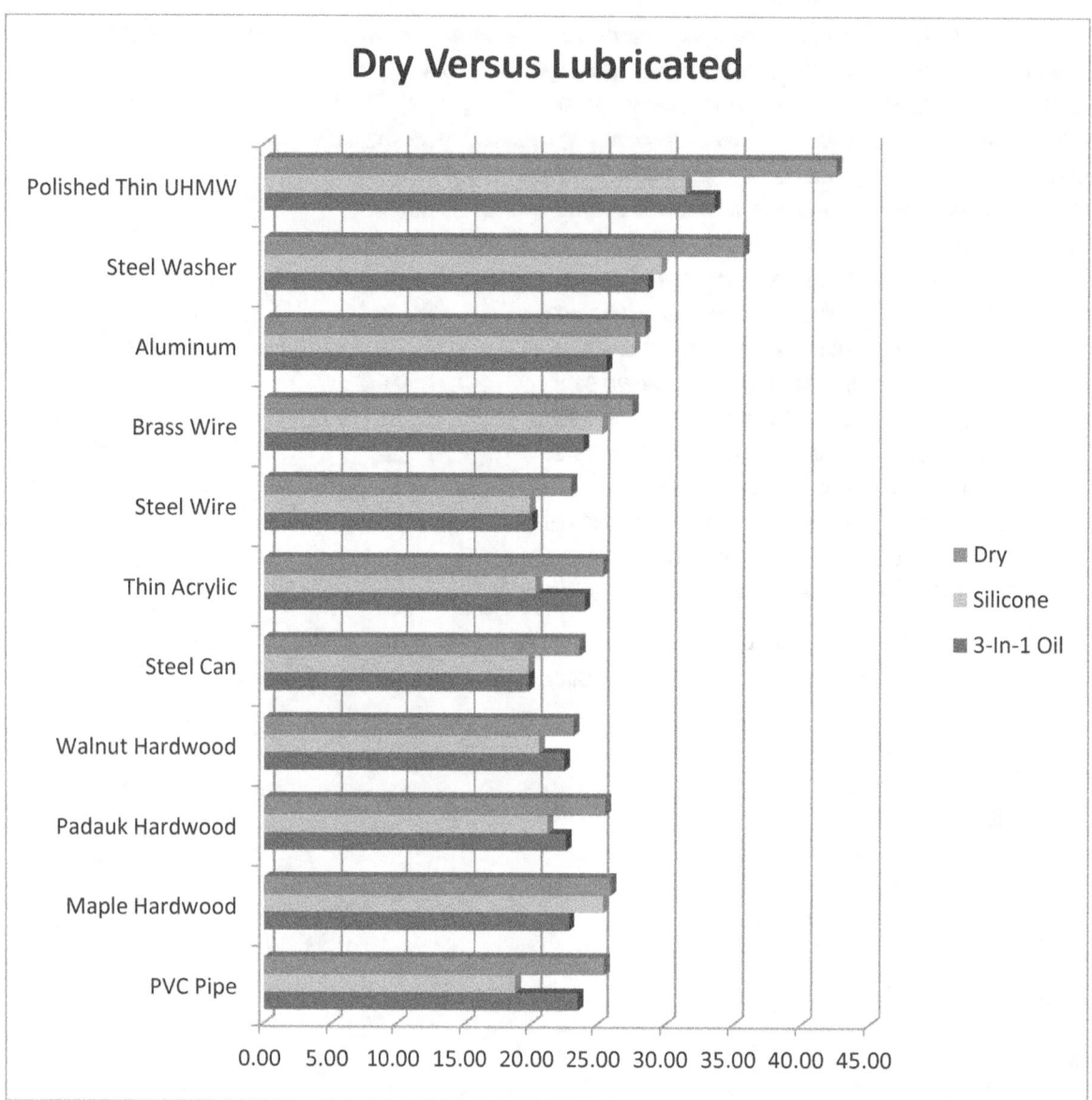

This is the data table for the comparison of average glide times with and without lubrication:

	Dry	Silicone	3-In-1 Oil
PVC Pipe	25.51	18.80	23.57
Maple Hardwood	25.97	25.41	22.87
Padauk Hardwood	25.57	21.20	22.66
Walnut Hardwood	23.25	20.56	22.52
Steel Can	23.70	19.79	19.79
Thin Acrylic	25.39	20.32	24.04
Steel Wire	23.03	19.87	20.02
Brass Wire	27.51	25.31	23.91
Aluminum	28.39	27.61	25.58
Steel Washer	35.65	29.61	28.63
Polished Thin UHMW	42.44	31.39	33.51